Kelley Wingate
Math Practice

Second Grade

Credits
Content Editor: Angela Triplett
Copy Editor: Christine Schwab

Visit *carsondellosa.com* for correlations to Common Core, state, national, and Canadian provincial standards.

Carson-Dellosa Publishing, LLC
PO Box 35665
Greensboro, NC 27425 USA
carsondellosa.com

ISBN 978-1-4838-0500-9

01-059141151

Table of Contents

Introduction

Competency in basic math skills creates a foundation for the successful use of math principles in the real world. Practicing math skills—in the areas of operations, algebra, place value, fractions, measurement, and geometry—is the best way to improve at them.

This book was developed to help students practice and master basic mathematical concepts. The practice pages can be used first to assess proficiency and later as basic skill practice. The extra practice will help students advance to more challenging math work with confidence. Help students catch up, stay up, and move ahead.

Common Core State Standards (CCSS) Alignment

This book supports standards-based instruction and is aligned to the CCSS. The standards are listed at the top of each page for easy reference. To help you meet instructional, remediation, and individualization goals, consult the Common Core State Standards alignment chart on page 4.

Leveled Activities

Instructional levels in this book vary. Each area of the book offers multilevel math activities so that learning can progress naturally. There are three levels, signified by one, two, or three dots at the bottom of the page:

- Level I: These activities will offer the most support.
- Level II: Some supportive measures are built in.
- Level III: Students will understand the concepts and be able to work independently.

All children learn at their own rate. Use your own judgment for introducing concepts to children when developmentally appropriate.

Hands-On Learning

Review is an important part of learning. It helps to ensure that skills are not only covered but are internalized. The flash cards at the back of this book will offer endless opportunities for review. Use them for a basic math facts drill, or to play bingo or other fun games.

There is also a certificate template at the back of this book for use as students excel at daily assignments or when they finish a unit.

Common Core State Standards Alignment Chart

Common Core State Standards*		Practice Page(s)
Operations and Algebraic Thinking		
Represent and solve problems involving addition and subtraction.	2.OA.1	5–13, 29
Add and subtract within 20.	2.OA.2	14–22
Work with equal groups of objects to gain foundations for multiplication.	2.OA.3–2.OA.4	23–28
Number and Operations in Base Ten		
Understand place value.	2.NBT.1–2.NBT.4	29–43, 68–70
Use place value understanding and properties of operations to add and subtract.	2.NBT.5–2NBT.9	44–70
Measurement and Data		
Measure and estimate lengths in standard units.	2.MD.1–2.MD.4	71–73
Relate addition and subtraction to length.	2.MD.5–2.MD.6	74–76
Work with time and money.	2.MD.7–2.MD.8	40, 77–85
Represent and interpret data.	2.MD.9–2.MD.10	86–91, 93
Geometry		
Reason with shapes and their attributes.	2.G.1–2.G.3	92–103

Name ______________________________

2.OA.A.1

Fact Families

Some addition and subtraction problems are related, like families. You can make two addition and two subtraction problems using the same three numbers.

$$5 + 6 = 11$$
$$6 + 5 = 11$$
$$11 - 5 = 6$$
$$11 - 6 = 5$$

Complete each fact family.

1. $2 + 8 = 10$ ____ $+ 2 = 10$ $10 - 8 = 2$ $10 - 2 =$ ____	2. $5 + 4 = 9$ $4 +$ ____ $= 9$ $9 - 5 =$ ____ $9 - 4 =$ ____	3. $6 + 9 = 15$ $9 + 6 =$ ____ $15 - 9 = 6$ ____ $- 6 = 9$
4. $5 + 7 = 12$ $7 +$ ____ $= 12$ ____ $- 5 = 7$ ____ $- 7 = 5$	5. $3 + 9 = 12$ ____ $+ 3 = 12$ $12 - 9 = 3$ ____ $- 3 =$ ____	6. $7 + 9 = 16$ $9 +$ ____ $= 16$ $16 -$ ____ $=$ ____ ____ $- 7 = 9$

Name ____________________

2.OA.A.1

Fact Families

Fact families use the same three numbers in addition and subtraction facts.

Complete each fact family.

1.

5 + 6 = ____

6 + 5 = ____

11 − 5 = ____

11 − 6 = ____

2.

4 + 8 = ____

8 + 4 = ____

12 − 8 = ____

12 − 4 = ____

3.

7 + 8 = ____

8 + 7 = ____

____ − 7 = ____

15 − ____ = ____

4.

____ + 8 = ____

8 + 9 = ____

17 − ____ = ____

17 − 9 = ____

5.

6 + 3 = ____

____ + 6 = ____

9 − 3 = ____

9 − ____ = ____

6.

7 + 6 = ____

____ + ____ = ____

13 − 7 = ____

____ − ____ = ____

Name ____________________ 2.OA.A.1

Fact Families

Complete each fact family.

1. 8 14 6	2. 5 12 7
___ + ___ = ___ ___ + ___ = ___ ___ − ___ = ___ ___ − ___ = ___	___ + ___ = ___ ___ + ___ = ___ ___ − ___ = ___ ___ − ___ = ___
3. 8 11 3	4. 7 15 8
___ + ___ = ___ ___ + ___ = ___ ___ − ___ = ___ ___ − ___ = ___	___ + ___ = ___ ___ + ___ = ___ ___ − ___ = ___ ___ − ___ = ___
5. 7 13 6	6. 9 17 8
___ + ___ = ___ ___ + ___ = ___ ___ − ___ = ___ ___ − ___ = ___	___ + ___ = ___ ___ + ___ = ___ ___ − ___ = ___ ___ − ___ = ___

Name ______________________

2.OA.A.1

Number Sentences

Each part of an addition problem has a name. The numbers being added are called **addends**, and the answer is called the **sum**. They form a number sentence. The addends can be switched around, and the sum stays the same! This is called the **commutative property of addition**.

Example: $8 + 7 = 15$ $7 + 8 = 15$ So . . . $8 + 7 = 7 + 8$

(8 and 7: addends; 15: sum)

Fill in the missing addends and sums.

1. $6 + \square = 10$
2. $5 + \square = 13$
3. $7 + \square = 12$
4. $\square + 7 = 14$
5. $\square + 9 = 11$
6. $\square + 8 = 10$
7. $8 + \square = 12$
8. $4 + \square = 10$
9. $9 + \square = 15$
10. $\square + 5 = 10$
11. $\square + 8 = 9$
12. $\square + 7 = 13$
13. $7 + \square = 11$
14. $6 + \square = 12$
15. $9 + \square = 14$
16. $8 + 7 = \square + 8$
17. $\square + 5 = 5 + 9$

Name ______________________ 2.OA.A.1

Number Sentences

Fill in the missing addends and sums.

1. 6 + ☐ = 16
2. 7 + ☐ = 13
3. 10 + ☐ = 12
4. ☐ + 7 = 11
5. ☐ + 9 = 18
6. ☐ + 8 = 17
7. 8 + ☐ = 18
8. 4 + ☐ = 13
9. 9 + ☐ = 14
10. ☐ + 5 = 15
11. ☐ + 8 = 14
12. ☐ + 7 = 16
13. 7 + ☐ = 16
14. 9 + ☐ = 12
15. 10 + ☐ = 14
16. ☐ + 9 = 17
17. ☐ + 8 = 11
18. ☐ + 7 = 15
19. 8 + 7 = ☐ + 8
20. ☐ + 7 = 7 + 9
21. 6 + ☐ = 9 + 6
22. 7 + 6 = 6 + ☐

Name ______________________________ (2.OA.A.1)

Number Sentences

Fill in the missing addends and sums.

1. 6 + □ = 18
2. 5 + □ = 16
3. 7 + □ = 19
4. □ + 7 = 17
5. □ + 9 = 20
6. □ + 8 = 20
7. 8 + □ = 15
8. 4 + □ = 13
9. 9 + □ = 19
10. □ + 5 = 20
11. □ + 8 = 16
12. □ + 7 = 20
13. 7 + □ = 19
14. 6 + □ = 20
15. 9 + □ = 18
16. □ + 9 = 18
17. □ + 8 = 11
18. □ + 7 = 14
19. 18 + 17 = □ + 18
20. □ + 15 = 15 + 9
21. 60 + □ = 90 + 60
22. 70 + 60 = 60 + □

Name ________________________________ 2.OA.A.1

Using Number Sentences in Word Problems

Use these steps to solve word problems:

1. Read the story and the question.
2. Read the question again.
3. Circle the numbers in the story that you need to answer the question.
4. Watch for key words: *altogether, how many more, in all.*
5. Choose to **+** or **−**.
6. Answer the question.
7. Use pictures, words, or numbers to show your work.

Solve each problem with a number sentence. Show your work with pictures, words, or numbers.

1. Megan has 10 baseball cards. Taylor has 14 baseball cards. How many more baseball cards does Taylor have than Megan?

   10 + _____ = 14 or 14 − 10 = _____ baseball cards

2. Matt has 9 crayons. Josh has 8 more crayons than Matt. How many crayons does Josh have?

3. Kevin has 11 more toy cars than Luke. Kevin has 16 toy cars. How many toy cars does Luke have?

4. A team has 12 students. Nine of the students are girls. How many students are boys?

5. A bus has 15 students riding on it. At the first bus stop, seven students get off. How many students are left on the bus?

Name ________________________ 2.OA.A.1

Using Number Sentences in Word Problems

Solve each problem with a number sentence. Show your work with pictures, words, or numbers.

1. Megan has 20 baseball cards. Taylor has 42 baseball cards. How many more baseball cards does Taylor have than Megan?

   20 + _____ = 42 or 42 − 20 = _____ baseball cards

2. Matt has 29 crayons. Josh has 11 more crayons than Matt. How many crayons does Josh have?

3. Kevin has 31 more toy cars than Luke. Kevin has 51 toy cars. How many toy cars does Luke have?

4. There are 26 students on a team. Fifteen of the students are girls. How many students are boys?

5. There are 39 students on a bus. Eighteen students get off the bus at the first stop. How many students are left on the bus?

6. There are 18 crackers on a plate. Miquel ate 4 crackers. Then, Nathan ate 2 crackers. How many crackers are left on the plate?

Name ____________________ 2.OA.A.1

Using Number Sentences in Word Problems

Solve the problem with a number sentence. Show your work with pictures, words, or numbers.

1. Megan has 76 baseball cards. Taylor has 89 baseball cards. How many more baseball cards does Taylor have than Megan?

2. Matt has 65 crayons. Josh has 33 more crayons than Matt. How many crayons does Josh have?

3. Kevin has 21 more toy cars than Luke. Kevin has 61 toy cars. How many toy cars does Luke have?

4. There are 45 students on a team. Twenty-three of the students are girls. How many students are boys?

5. There are 69 students on a bus. Seventeen students get off the bus at the first stop. How many students are left on the bus?

6. There are 32 crackers on a plate. Miquel ate 16 crackers. Then, Nathan ate 7 crackers. How many crackers are left on the plate?

Name ______________________________

2.OA.B.2

Addition Fluency

Solve each problem.

1. $\begin{array}{r} 7 \\ +\ 6 \\ \hline \end{array}$	2. $\begin{array}{r} 7 \\ +\ 7 \\ \hline \end{array}$	3. $\begin{array}{r} 3 \\ +\ 6 \\ \hline \end{array}$	4. $\begin{array}{r} 4 \\ +\ 3 \\ \hline \end{array}$	5. $\begin{array}{r} 7 \\ +\ 4 \\ \hline \end{array}$	6. $\begin{array}{r} 5 \\ +\ 4 \\ \hline \end{array}$
7. $\begin{array}{r} 6 \\ +\ 5 \\ \hline \end{array}$	8. $\begin{array}{r} 9 \\ +\ 2 \\ \hline \end{array}$	9. $\begin{array}{r} 8 \\ +\ 4 \\ \hline \end{array}$	10. $\begin{array}{r} 6 \\ +\ 2 \\ \hline \end{array}$	11. $\begin{array}{r} 3 \\ +\ 2 \\ \hline \end{array}$	12. $\begin{array}{r} 3 \\ +\ 9 \\ \hline \end{array}$
13. $\begin{array}{r} 3 \\ +\ 3 \\ \hline \end{array}$	14. $\begin{array}{r} 9 \\ +\ 5 \\ \hline \end{array}$	15. $\begin{array}{r} 9 \\ +\ 6 \\ \hline \end{array}$	16. $\begin{array}{r} 2 \\ +\ 7 \\ \hline \end{array}$	17. $\begin{array}{r} 6 \\ +\ 4 \\ \hline \end{array}$	18. $\begin{array}{r} 5 \\ +\ 8 \\ \hline \end{array}$
19. $\begin{array}{r} 3 \\ +\ 7 \\ \hline \end{array}$	20. $\begin{array}{r} 6 \\ +\ 8 \\ \hline \end{array}$	21. $\begin{array}{r} 8 \\ +\ 3 \\ \hline \end{array}$	22. $\begin{array}{r} 5 \\ +\ 4 \\ \hline \end{array}$	23. $\begin{array}{r} 9 \\ +\ 1 \\ \hline \end{array}$	24. $\begin{array}{r} 2 \\ +\ 8 \\ \hline \end{array}$
25. $\begin{array}{r} 2 \\ +\ 3 \\ \hline \end{array}$	26. $\begin{array}{r} 3 \\ +\ 5 \\ \hline \end{array}$	27. $\begin{array}{r} 4 \\ +\ 9 \\ \hline \end{array}$	28. $\begin{array}{r} 4 \\ +\ 3 \\ \hline \end{array}$	29. $\begin{array}{r} 8 \\ +\ 8 \\ \hline \end{array}$	30. $\begin{array}{r} 3 \\ +\ 3 \\ \hline \end{array}$

Name ______________________________ 2.OA.B.2

Addition Fluency

Solve each problem.

1. $6 + 3$
2. $7 + 3$
3. $5 + 5$
4. $8 + 2$
5. $8 + 3$
6. $6 + 4$
7. $5 + 3$
8. $3 + 4$
9. $6 + 6$
10. $3 + 3$
11. $9 + 0$
12. $7 + 3$
13. $9 + 3$
14. $9 + 2$
15. $5 + 4$
16. $1 + 2$
17. $0 + 9$
18. $8 + 4$
19. $4 + 4$
20. $7 + 5$
21. $3 + 8$
22. $8 + 4$
23. $3 + 7$
24. $7 + 2$
25. $6 + 2$
26. $5 + 6$
27. $5 + 1$
28. $4 + 5$
29. $9 + 2$
30. $6 + 6$

Name ______________________________ 2.OA.B.2

Addition Fluency

Solve each problem.

1.	2.	3.	4.	5.	6.
9 + 5	7 + 7	8 + 5	6 + 7	5 + 5	6 + 4

7.	8.	9.	10.	11.	12.
9 + 7	9 + 4	5 + 6	9 + 9	6 + 3	7 + 4

13.	14.	15.	16.	17.	18.
8 + 6	7 + 5	9 + 7	9 + 5	8 + 7	9 + 3

19.	20.	21.	22.	23.	24.
8 + 2	9 + 4	8 + 8	9 + 3	8 + 3	9 + 8

25.	26.	27.	28.	29.	30.
6 + 6	7 + 3	9 + 9	14 + 2	16 + 2	17 + 1

Name ____________________ 2.OA.B.2

Subtraction Fluency

Solve each problem.

1. $9 - 6$
2. $8 - 2$
3. $10 - 10$
4. $10 - 6$
5. $10 - 7$
6. $9 - 6$
7. $7 - 2$
8. $7 - 4$
9. $10 - 4$
10. $10 - 0$
11. $9 - 4$
12. $10 - 2$
13. $8 - 6$
14. $10 - 2$
15. $7 - 5$
16. $9 - 5$
17. $8 - 5$
18. $8 - 7$
19. $7 - 3$
20. $8 - 4$
21. $10 - 3$
22. $6 - 2$
23. $10 - 1$
24. $9 - 3$
25. $7 - 6$
26. $10 - 4$
27. $10 - 8$
28. $10 - 5$
29. $8 - 1$
30. $10 - 9$

Name ______________________________ 2.OA.B.2

Subtraction Fluency

Solve each problem.

1. $10 - 9$	2. $15 - 9$	3. $13 - 5$	4. $15 - 8$	5. $12 - 7$	6. $14 - 6$
7. $10 - 4$	8. $12 - 8$	9. $13 - 9$	10. $15 - 5$	11. $12 - 5$	12. $15 - 3$
13. $14 - 5$	14. $12 - 6$	15. $12 - 4$	16. $11 - 6$	17. $15 - 2$	18. $10 - 6$
19. $13 - 6$	20. $13 - 8$	21. $12 - 9$	22. $13 - 3$	23. $15 - 7$	24. $14 - 7$
25. $15 - 8$	26. $10 - 5$	27. $15 - 1$	28. $12 - 1$	29. $15 - 2$	30. $12 - 6$

Name ______________________________

2.OA.B.2

Subtraction Fluency

Solve each problem.

1. $10 - 5$	2. $15 - 6$	3. $12 - 8$	4. $12 - 6$	5. $13 - 7$	6. $12 - 5$
7. $17 - 9$	8. $11 - 8$	9. $11 - 9$	10. $18 - 9$	11. $14 - 5$	12. $14 - 9$
13. $14 - 6$	14. $13 - 9$	15. $14 - 8$	16. $11 - 3$	17. $16 - 8$	18. $15 - 8$
19. $10 - 2$	20. $16 - 7$	21. $15 - 9$	22. $11 - 2$	23. $16 - 9$	24. $15 - 7$
25. $13 - 4$	26. $13 - 8$	27. $10 - 8$	28. $18 - 2$	29. $17 - 1$	30. $12 - 9$

Name ______________________________ 2.OA.B.2

Addition and Subtraction Fluency

Solve each problem.

1. 6 + 3 =
2. 8 − 6 =
3. 5 + 3 =
4. 10 − 10 =
5. 9 + 0 =
6. 9 − 0 =
7. 8 + 1 =
8. 8 + 3 =
9. 7 − 2 =
10. 5 + 1 =
11. 9 − 2 =
12. 5 + 4 =
13. 6 − 4 =
14. 7 + 3 =
15. 8 − 2 =
16. 2 + 6 =
17. 10 − 6 =
18. 10 − 1 =
19. 6 + 4 =
20. 8 − 2 =
21. 2 + 7 =
22. 8 − 5 =
23. 5 + 5 =
24. 5 − 5 =

Name ______________________________ (2.OA.B.2)

Addition and Subtraction Fluency

Solve each problem.

1. 18 − 9 =	2. 10 + 5 =	3. 13 − 7 =
4. 14 + 6 =	5. 15 − 6 =	6. 12 + 3 =
7. 15 − 7 =	8. 14 + 1 =	9. 17 − 8 =
10. 11 − 7 =	11. 13 − 4 =	12. 11 + 4 =
13. 12 + 6 =	14. 15 − 9 =	15. 11 − 5 =
16. 10 − 1 =	17. 14 + 5 =	18. 16 − 8 =
19. 10 − 2 =	20. 13 − 5 =	21. 10 + 8 =
22. 11 + 2 =	23. 10 + 9 =	24. 11 − 8 =
25. 13 − 8 =	26. 17 − 9 =	27. 16 + 2 =
28. 16 + 1 =	29. 18 − 0 =	30. 15 + 5 =

Name ______________________________ 2.OA.B.2

Addition and Subtraction Fluency

Solve each problem.

1. 13 − 8 =
2. 15 − 6 =
3. 17 − 8 =
4. 16 − 9 =
5. 18 − 9 =
6. 14 − 9 =
7. 13 − 9 =
8. 13 − 6 =
9. 12 + 5 =
10. 10 + 3 =
11. 12 − 9 =
12. 15 − 9 =
13. 15 − 8 =
14. 11 + 9 =
15. 15 − 8 =
16. 11 − 8 =
17. 16 − 7 =
18. 10 + 4 =
19. 15 + 4 =
20. 12 + 7 =
21. 17 − 9 =
22. 10 + 3 =
23. 16 − 8 =
24. 16 − 9 =
25. 12 + 3 =
26. 14 − 6 =
27. 11 + 7 =
28. 12 + 4 =
29. 18 + 0 =
30. 17 − 9 =

Name ______________________________ 2.OA.C.3

Odd and Even

When counting groups of objects that can be paired together evenly, we call the sum **even**. When there is one left over in the group, we call the sum **odd**. Even numbers end in 0, 2, 4, 6, and 8. Odd numbers end in 1, 3, 5, 7, and 9.

Circle groups of two in each problem. If the sum of the objects is an even number, write **E** in the box. If the sum of the objects is an odd number, write **O** in the box.

1. O	2.
3.	4.
5.	6.
7.	8.
9.	10.
11.	12.

Name ______________________________ (2.OA.C.3)

Odd and Even

Even numbers end in 0, 2, 4, 6, and 8.
Odd numbers end in 1, 3, 5, 7, and 9.

Count the candles on each cake. Write the number on the cake. Color each cake with an even number.

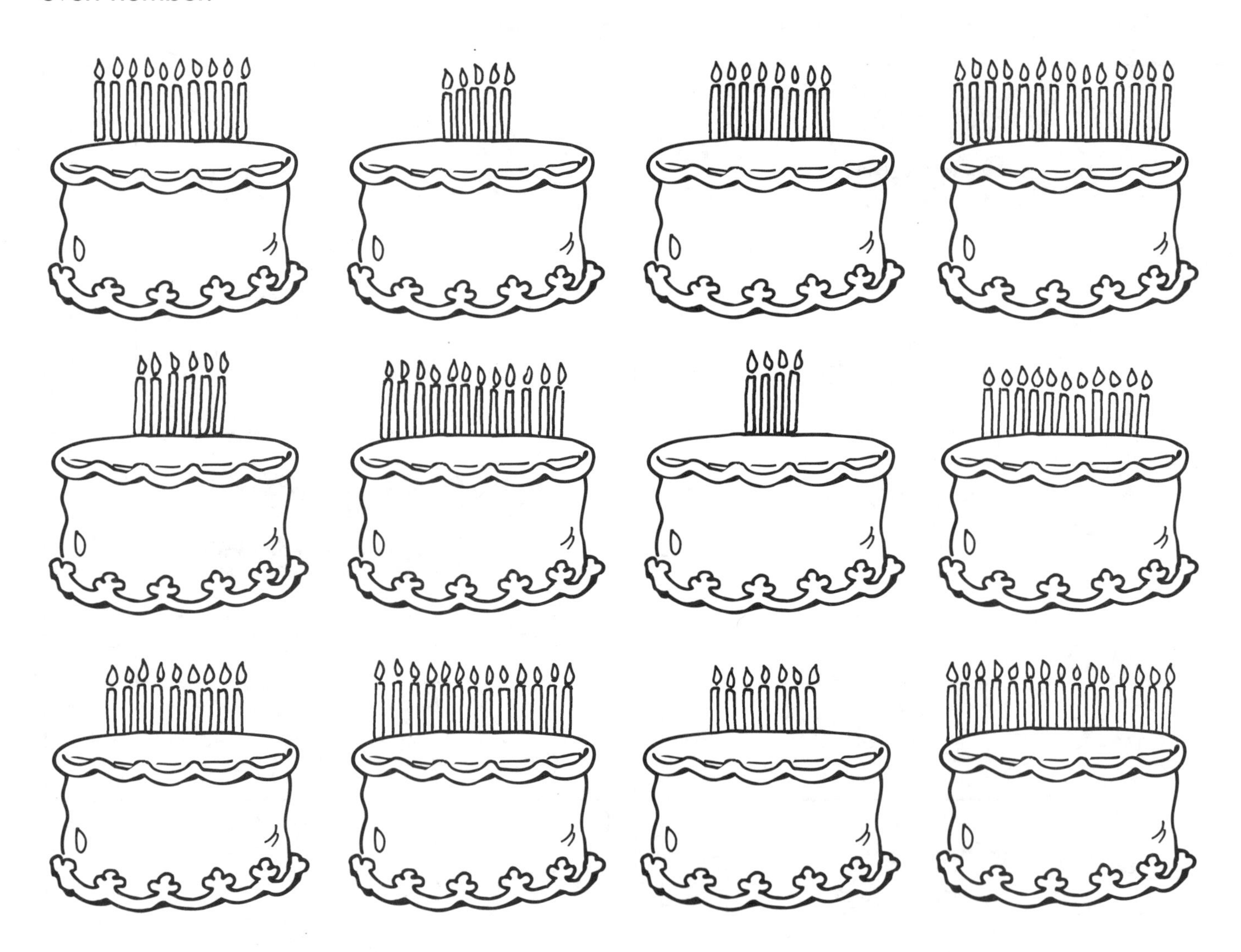

Draw candles on the cake to show how old you are.

Is your age even or odd? ______________

Name ____________________ 2.OA.C.3

Odd and Even

Draw a picture to go with each story. Then, count to find the total. Is the total odd or even?

1. You see two beach towels. On each towel, there are 4 buckets. How many buckets are there in all?  = 8 even	2. Waves washed four starfish onto the sand. Each starfish has 5 legs. How many legs is that altogether? = ☐ ______
3. In the ocean, 3 children are swimming. Each child is wearing 2 fins. How many fins are there combined? = ☐ ______	4. Near the shore, 5 children are playing in the waves. Each child has 3 balls. How many balls are there? = ☐ ______
5. There are 3 buckets on the beach. Each bucket has 4 shovels in it. How many shovels are there in all? = ☐ ______	6. The waves washed 7 seashells onto the sand. Each shell has an animal living in it. How many animals are there? = ☐ ______

Name ____________________ 2.OA.C.4

Using Arrays to Add Equal Groups

Objects placed in equal rows and columns are called **arrays**. A row goes across and a column goes down. Use repeated addition to find the sum of the objects in an array.

☆☆☆☆
☆☆☆☆ 2 + 2 + 2 + 2 = **8**

Add the objects in each column to find the total.

1. 2 + 2 + 2 = ____

2. 5 + 5 + 5 + 5 = ____

3. ____ + ____ + ____ + ____ + ____ + ____ = ____

4. ____ + ____ + ____ + ____ + ____ = ____

5. ____ + ____ + ____ = ____

6. ____ + ____ + ____ + ____ = ____

Name ______________________________ 2.OA.C.4

Using Arrays to Add Equal Groups

Add the objects in each column to find the total. Write a number sentence for each array.

1. ××× ××× ××× ______________________	2. ○○○○ ○○○○ ○○○○ ______________________
3. ☆☆ ☆☆ ☆☆ ______________________	4. @@@@@ @@@@@ @@@@@ ______________________
5. ××××× ______________________	6. ○○○○○ ○○○○○ ○○○○○ ○○○○○ ______________________
7. ☆☆ ☆☆ ______________________	8. @@@ @@@ @@@ @@@ @@@ ______________________
9. ××××× ××××× ××××× ××××× ××××× ______________________	10. ○○○○ ○○○○ ○○○○ ○○○○ ______________________

Name ______________________________ 2.OA.C.4

Using Arrays to Add Equal Groups

Add the columns to find the total number of squares in each rectangle. Write a repeated addition sentence for each array. Then, write a multiplication sentence.

1.	2.	3.	4.
______________ ______________	______________ ______________	______________ ______________	______________ ______________
5.	6.	7.	8.
______________ ______________	______________ ______________	______________ ______________	______________ ______________

Name ______________________________

2.NBT.A.1, 2.NBT.A.3

Place Value

Putting 10 **tens** together makes 1 **hundred**.

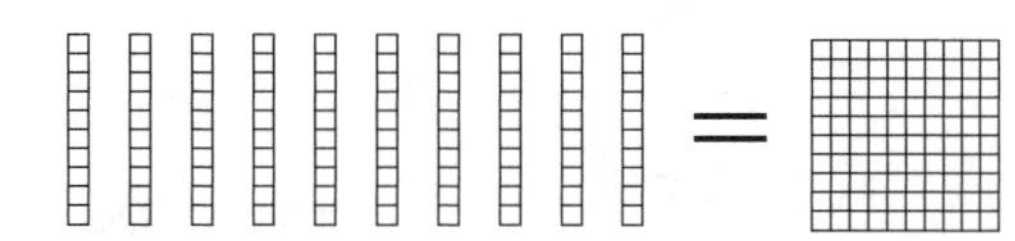

Count each group of blocks. Write the number of hundreds, tens, and ones.

1. 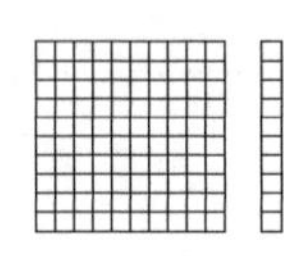 = hundreds _____ tens _____ ones _____

2. = hundreds _____ tens _____ ones _____

3. = hundreds _____ tens _____ ones _____

4. 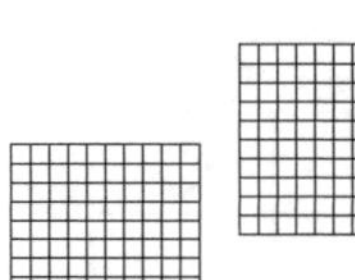 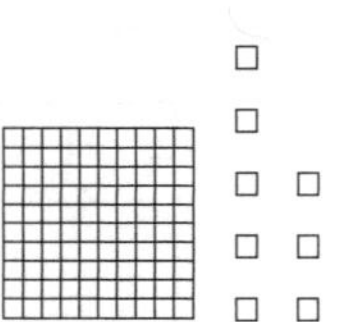 = hundreds _____ tens _____ ones _____

5. 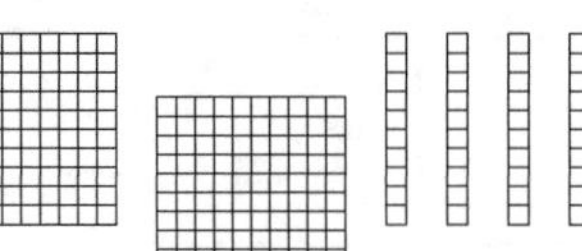 = hundreds _____ tens _____ ones _____

Name ______________________________ (2.NBT.A.1, 2.NBT.A.3)

Place Value

Imagine taking 10 tens and gluing them together like this:

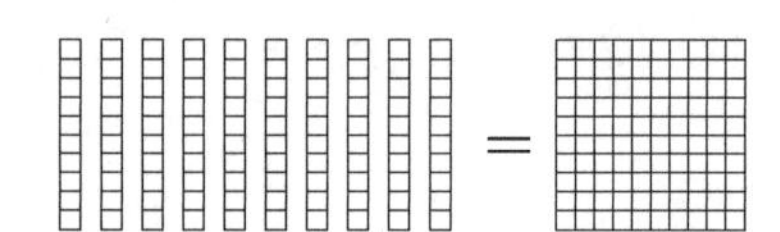

The block is now called a **hundred**. The ten sticks are still **tens**, and the leftover blocks are still **ones**.

hundred ten one

Count the hundreds, tens, and ones. Write the number two ways.

1.  =

H	T	O

= ______

2. 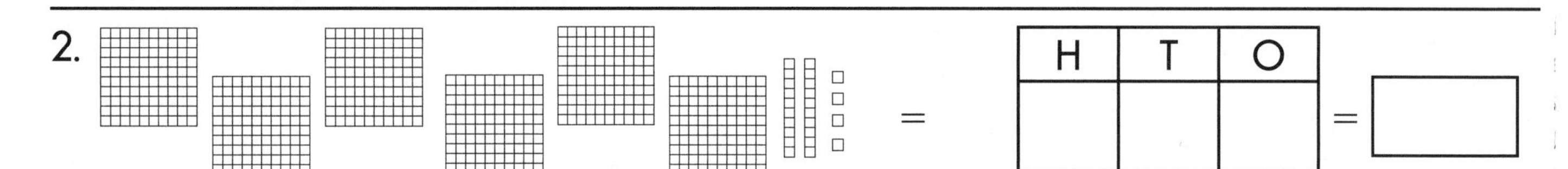 =

H	T	O

= ______

3.  =

H	T	O

= ______

4. 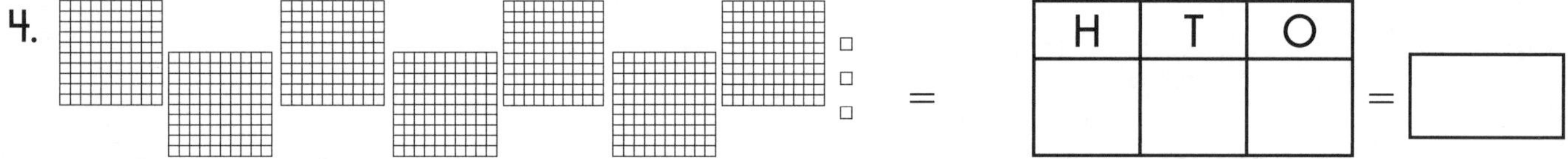 =

H	T	O

= ______

5. =

H	T	O

= ______

6.  =

H	T	O

= ______

7. =

H	T	O

= ______

Name ______________________________ 2.NBT.A.1, 2.NBT.A.3

Place Value

When you see a three-digit number, remember that you are looking at a hundreds place, tens place, and ones place.

Write each number another way.

Then, draw hundreds blocks 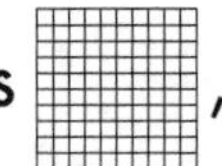, tens blocks, and ones blocks □.

1.
402 = _____ hundreds _____ tens _____ ones =

2.
523 = _____ hundreds _____ tens _____ ones =

3.
630 = _____ hundreds _____ tens _____ ones =

4.
241 = _____ hundreds _____ tens _____ ones =

5. Choose a three-digit number. On another sheet of paper, draw the number with place value blocks.

Name ______________________________ 2.NBT.A.2

Skip Counting

Skip counting means following a given pattern as you count. You can skip count by 5s, 10s and 100s.

Example:

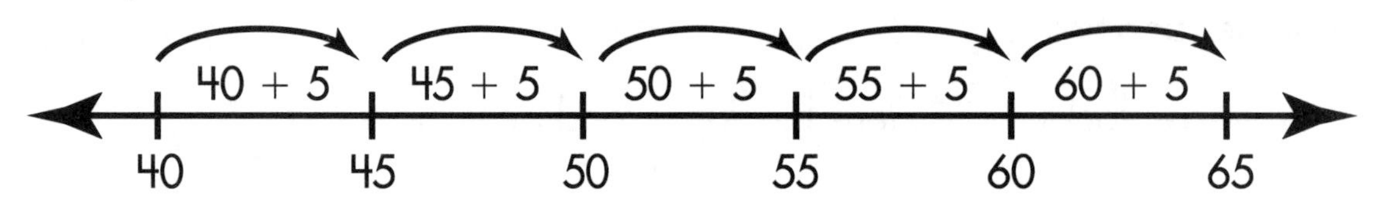

Look for a pattern. Write the missing numbers. Skip count by 5s.

1. 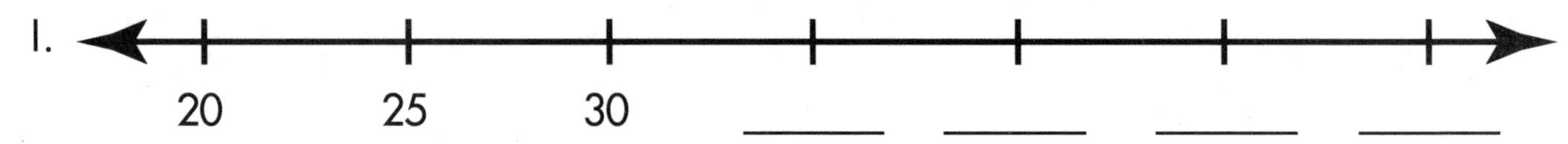

2. 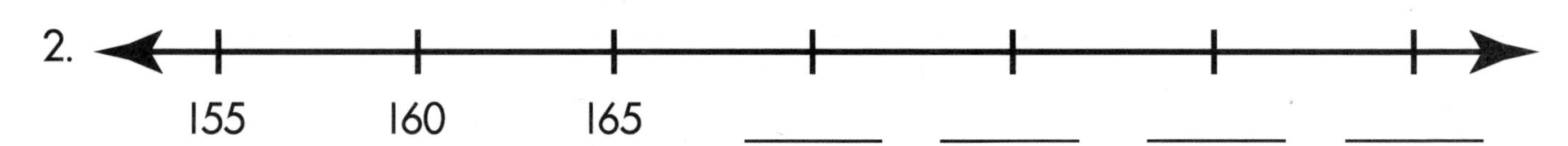

Skip count by 10s.

3. 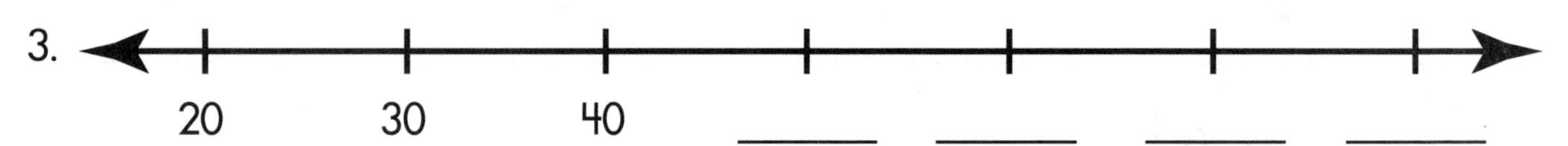

4. 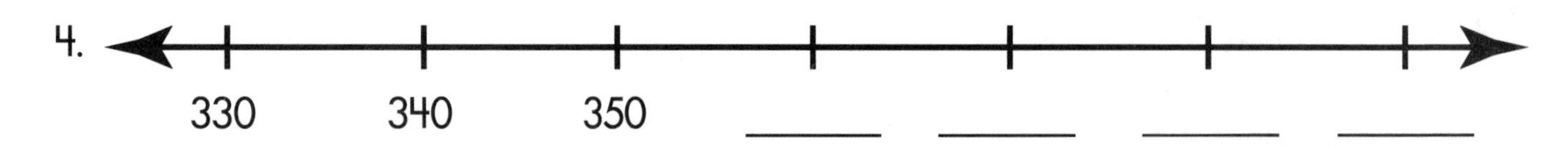

Skip count by 100s.

5. 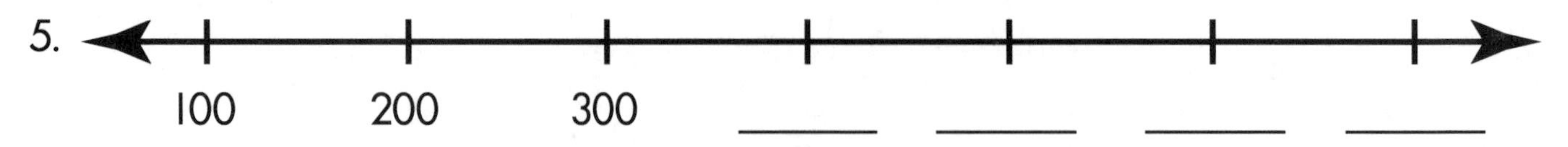

6. 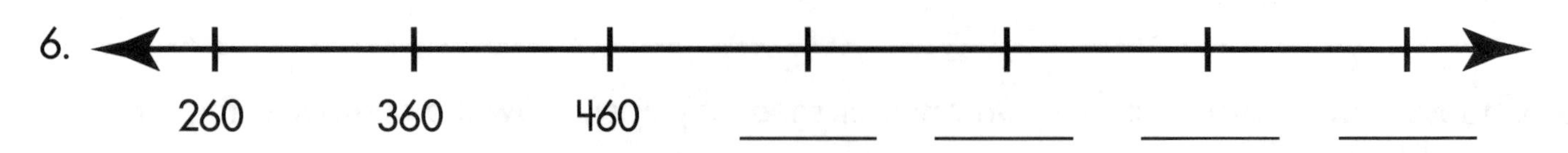

Name ______________________ (2.NBT.A.2)

Skip Counting

Skip count by 5s.

1.
2.

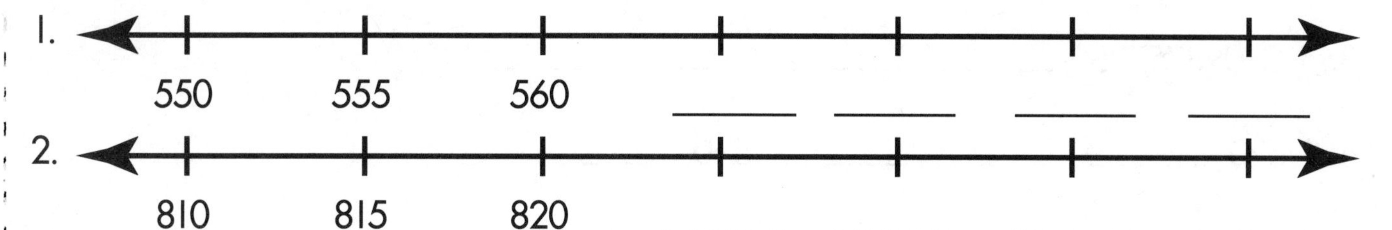

3. 245, 250, 255, 260, ______, ______, ______, ______

4. 505, 510, 515, 520, ______, ______, ______, ______

Skip count by 10s.

5.
6.

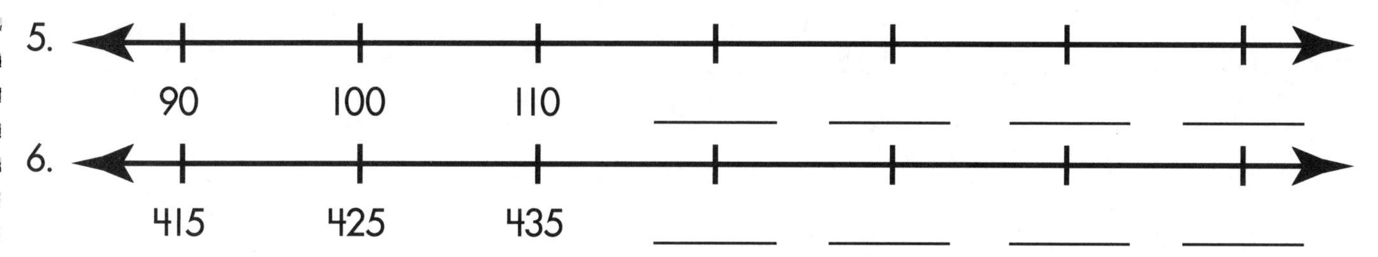

7. 310, 320, 330, 340, ______, ______, ______, ______

8. 635, 645, 655, 665, ______, ______, ______, ______

Skip count by 100s.

9.
10.

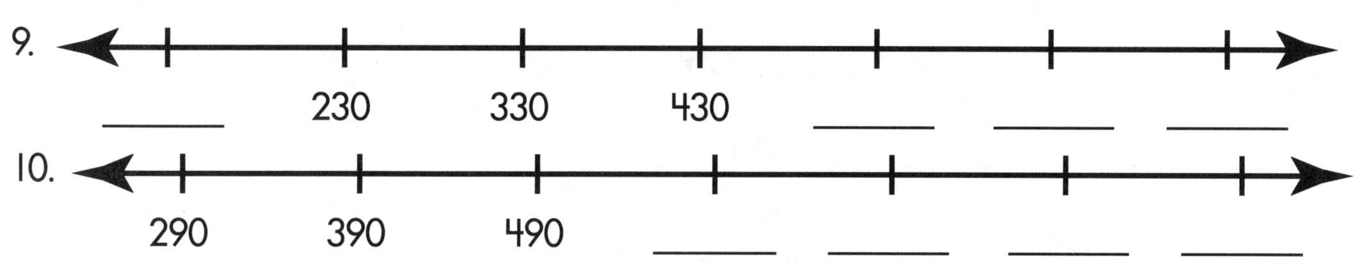

11. 195, 295, 395, 495, ______, ______, ______, ______

12. 224, 324, 424, 524, ______, ______, ______, ______

Name ______________________________ 2.NBT.A.2

Skip Counting

Skip count by 5s.

1. 365, 370, 375, 380, ______ , ______ , ______ , ______

2. 515, 520, ______ , ______ , ______ , 540, ______

3. 865, ______ , 855, ______ , 845, ______ , 835, ______

4. 770, 775, 780, ______ , 790, ______ , ______ , ______

Skip count by 10s.

5. 495, 505, ______ , ______ , 535, ______ , ______

6. ______ , 323, 333, 343, ______ , ______ , ______

7. 810, ______ , ______ , ______ , 850, ______

8. 488, ______ , ______ , 458, ______ , ______

Skip count by 100s.

9. 179, 279, ______ , ______ , ______ , 679, ______

10. ______ , 573, ______ , ______ , 873, ______

11. 915, ______ , 715, ______ , 515, ______ , 315

12. ______ , 299, ______ , ______ , ______ , 699

Name ______________________________ 2.NBT.A.1, 2.NBT.A.3

Reading and Writing Numbers in Expanded Notation

Three-digit numbers have three parts: the **hundreds** column (place), the **tens** column (place), and the **ones** column (place). The number can also be shown using blocks:

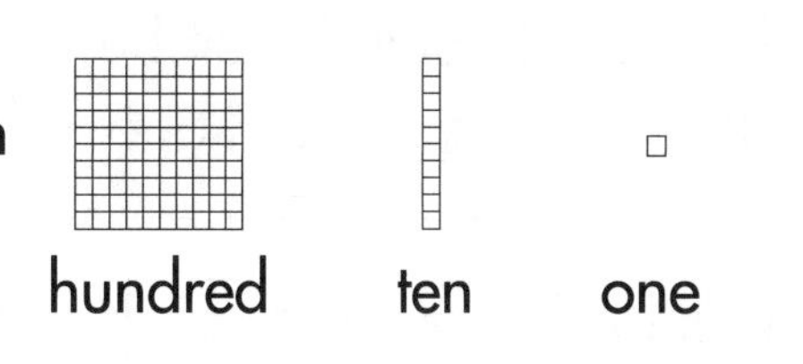

Count the hundreds, tens, and ones. Write the number two ways.

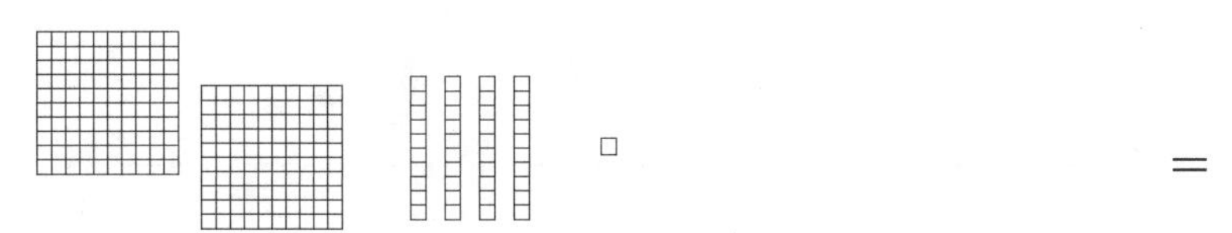

=

H	T	O
2	**4**	**1**

= **241**

1. 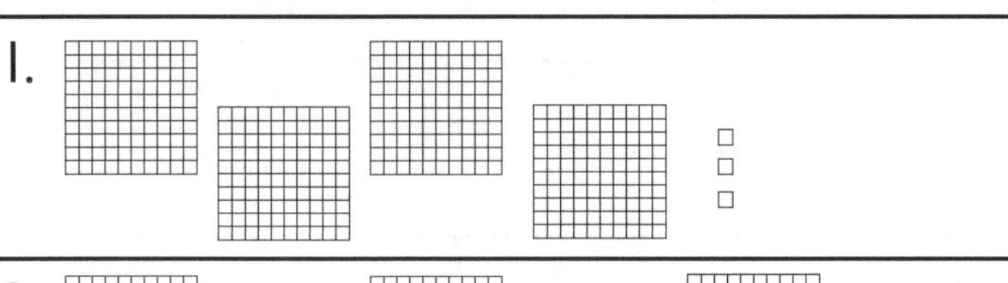

=

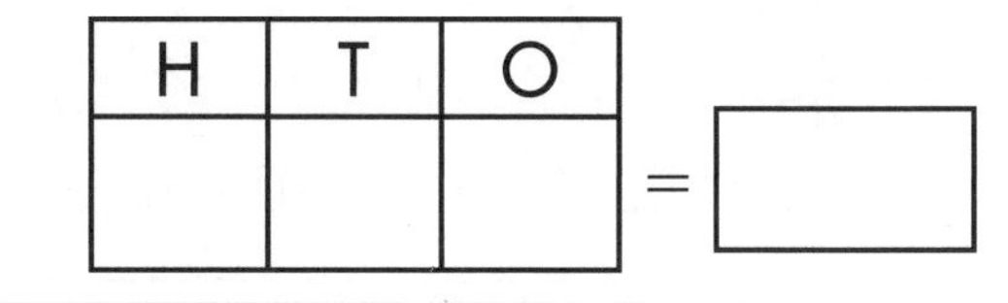

H	T	O

= ______

2. 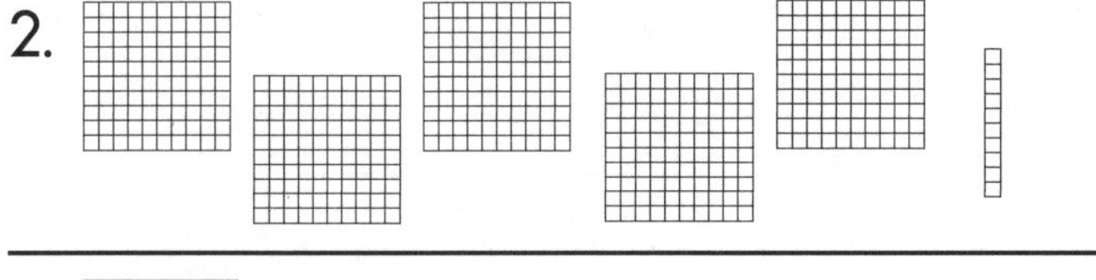

=

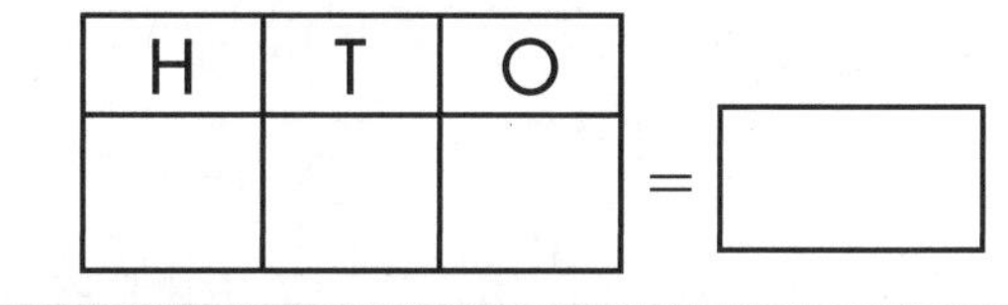

H	T	O

= ______

3. 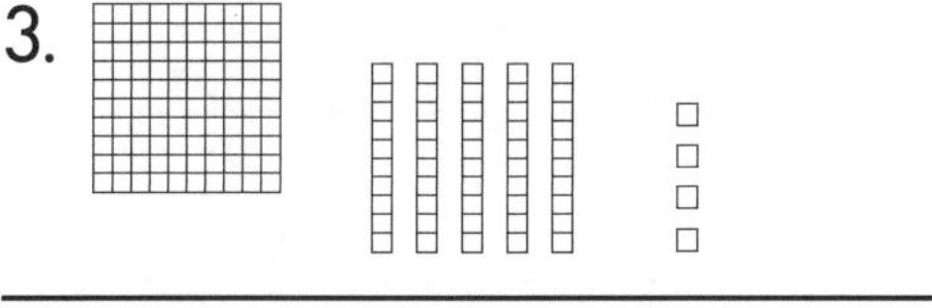

=

H	T	O

= ______

4. 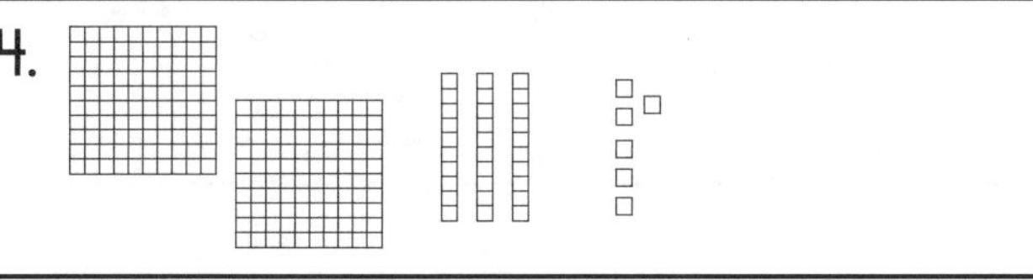

=

H	T	O

= ______

5. 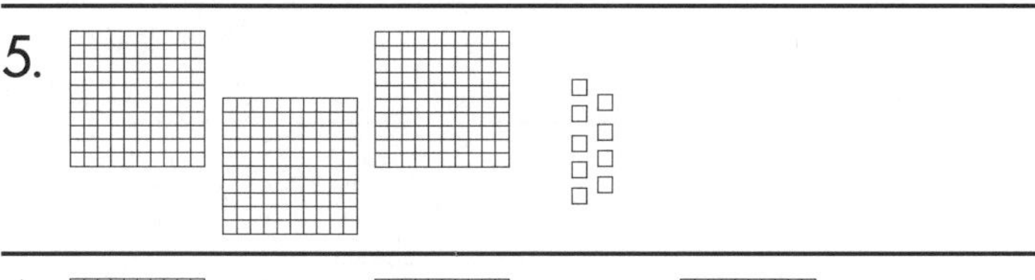

=

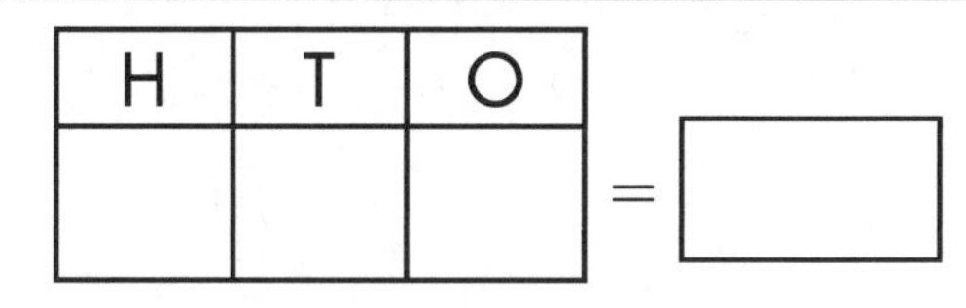

H	T	O

= ______

6. 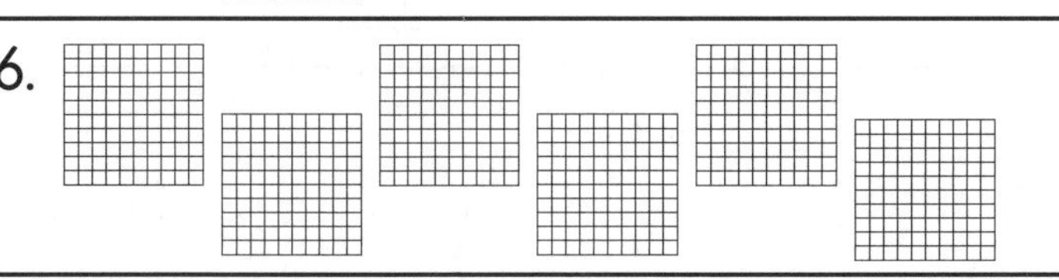

=

H	T	O

= ______

7.

=

H	T	O

= ______

Name ______________________________ (2.NBT.A.1, 2.NBT.A.3)

Reading and Writing Numbers in Expanded Notation

A three-digit number can be written as an addition problem by separating the hundreds, tens, and ones. This is called **expanded notation**.

Write an addition problem for each picture as shown. Then, write the number.

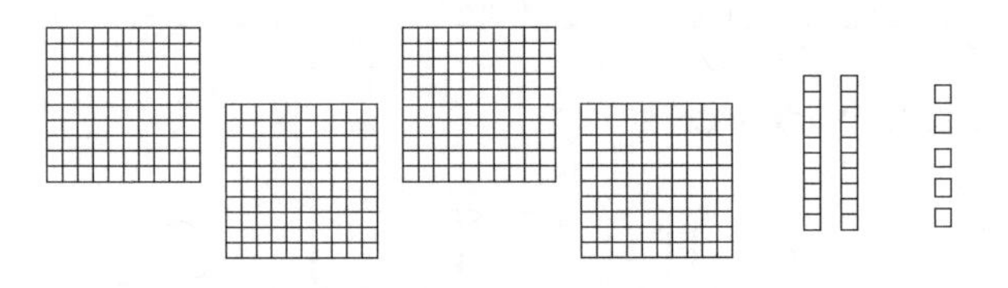

400 + **20** + **5**

425

1. 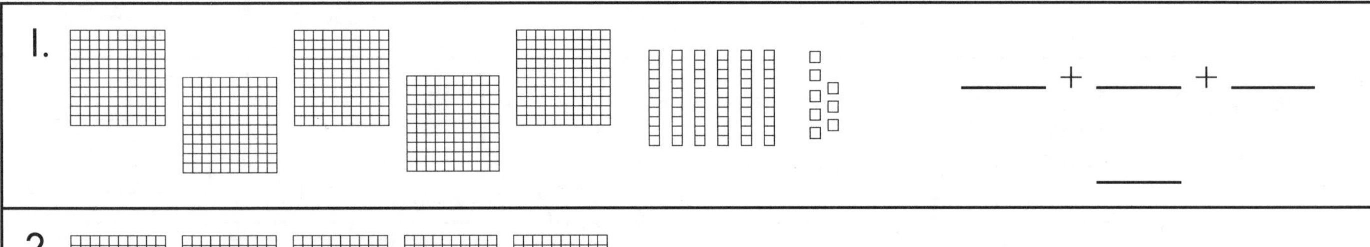

____ + ____ + ____

____

2. 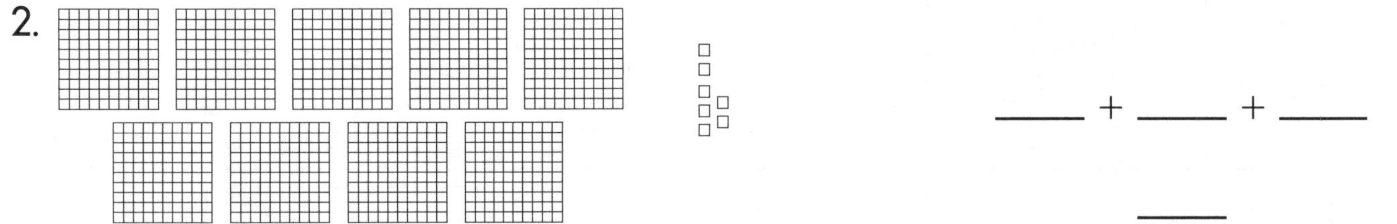

____ + ____ + ____

____

3. 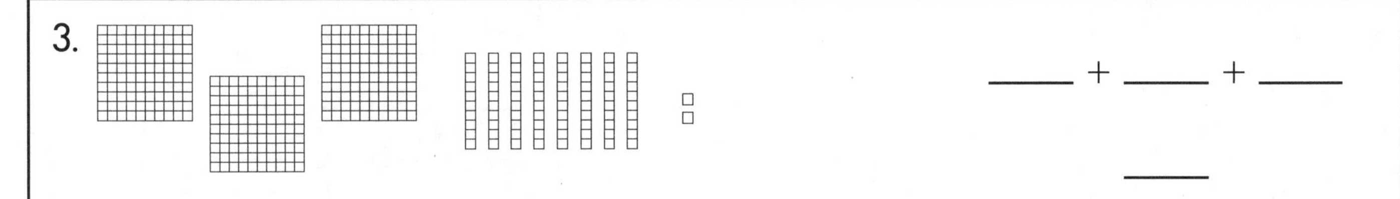

____ + ____ + ____

____

4. 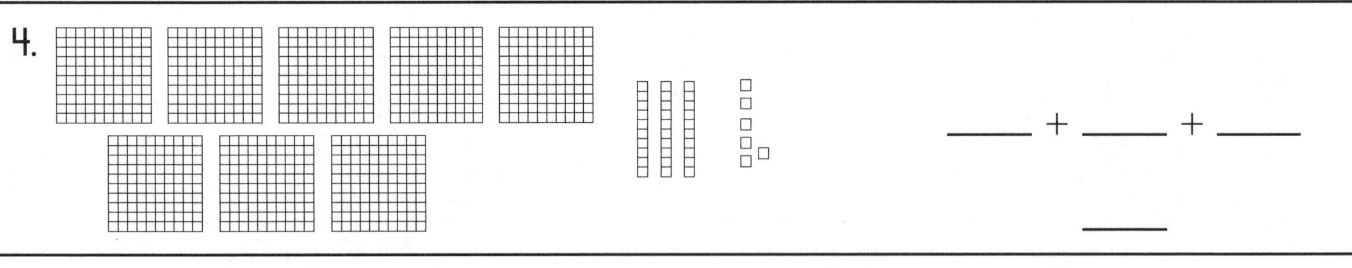

____ + ____ + ____

____

5. 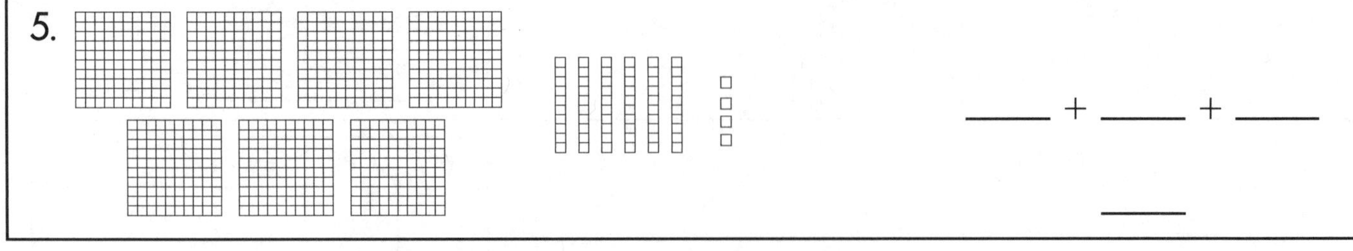

____ + ____ + ____

____

Name ______________________________ (2.NBT.A.1, 2.NBT.A.3)

Reading and Writing Numbers in Expanded Notation

To find the correct order of three-digit numbers, always look at the hundreds place first. The number with the most hundreds is the greatest number. If the numbers in the hundreds place are the same, go to the tens place. If they are the same, go to the ones. The higher number may be called **greatest**, **greater**, or **highest**. The lower number may be called **least**, **less**, or **lowest**.

Write the matching number in each box. Then, sequence the numbers from least to greatest on the lines at the bottom.

1. 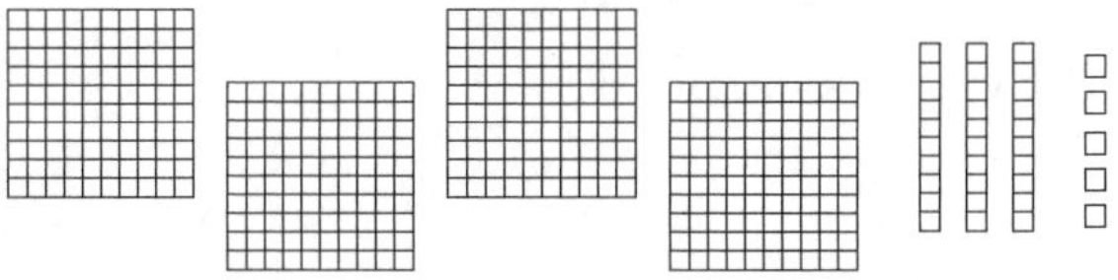	2. 7 hundreds 6 tens 8 ones
3. nine hundred ten	4. 300 + 70 + 8
5. 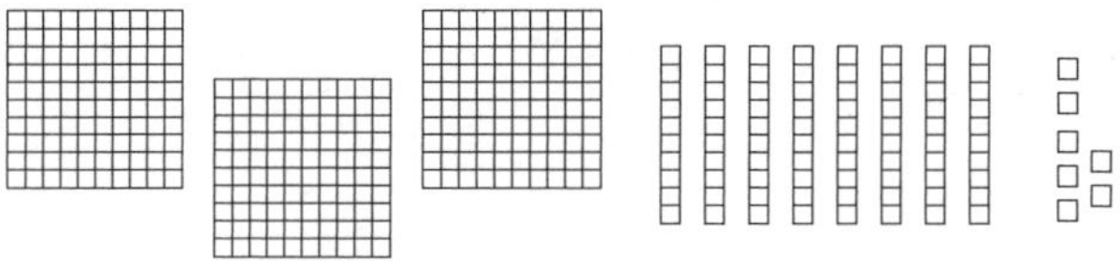	6. one hundred ninety-nine
7. 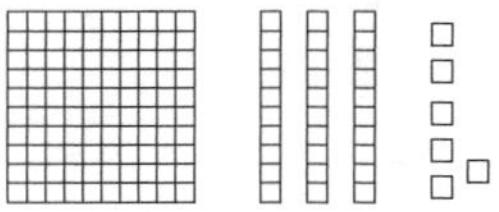	8. 6 hundreds 8 tens 3 ones
9. 600 + 10 + 4	10. eight hundred fifteen

Least Greatest

11. 136, ______, ______, ______, ______, ______, ______, ______, ______, ______

Name ______________________________ (2.NBT.A.4)

Comparing Numbers

Comparing numbers means deciding which number is greater and which number is smaller. The symbol > means **greater than** and the symbol < means **less than**. Use these steps to compare:

1. Are the number of digits the same? If not, the number with the most digits is larger.
2. If the number of digits is the same, begin with the digit on the left. Which number has a larger digit? That is the greater number.
3. If the digits are the same, move to the next place and find the larger number. If the numbers are the same, they are equal (=).

671 (<) 2,318	564 (>) 372	671 (>) 619
This number has more digits, so it is greater.	5 is greater.	Same, so go to the next digit. 7 is greater than 1.

Write **>**, **<**, or **=** to compare the numbers.

1. 17 ◯ 98
2. 298 ◯ 300
3. 82 ◯ 18
4. 761 ◯ 760
5. 79 ◯ 23
6. 395 ◯ 217
7. 76 ◯ 76
8. 514 ◯ 514
9. 63 ◯ 50
10. 330 ◯ 630
11. 102 ◯ 102
12. 129 ◯ 130

Name ______________________________ (2.NBT.A.4)

Comparing Numbers

Write **>**, **<**, or **=** to compare the numbers.

1. 886 ◯ 542
2. 130 ◯ 13
3. 119 ◯ 109
4. 903 ◯ 309
5. 984 ◯ 984
6. 153 ◯ 153
7. 578 ◯ 587
8. 600 ◯ 300
9. 907 ◯ 709
10. 999 ◯ 799
11. 534 ◯ 990
12. 865 ◯ 568
13. 760 ◯ 760
14. 712 ◯ 233
15. 521 ◯ 125
16. 222 ◯ 122
17. 18 ◯ 189
18. 905 ◯ 905
19. Explain how you know that 980 is greater than 880.

____

____

Name ______________________ (2.NBT.A.4, 2.MD.C.8)

Comparing Numbers

Find the value of each group of coins. Compare the values and Write **>**, **<**, or **=** in the circle.

1. ______¢

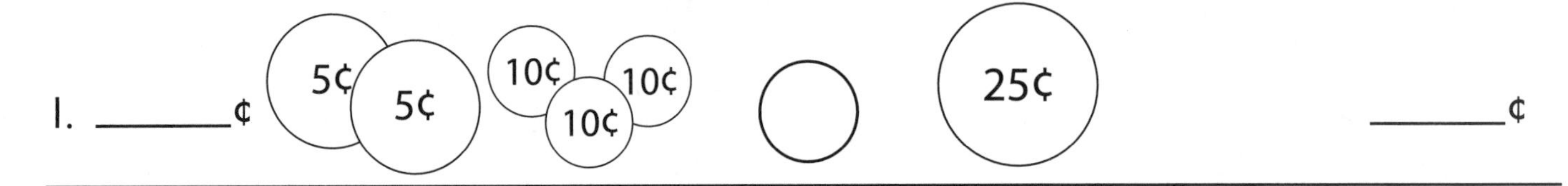

______¢

2. ______¢

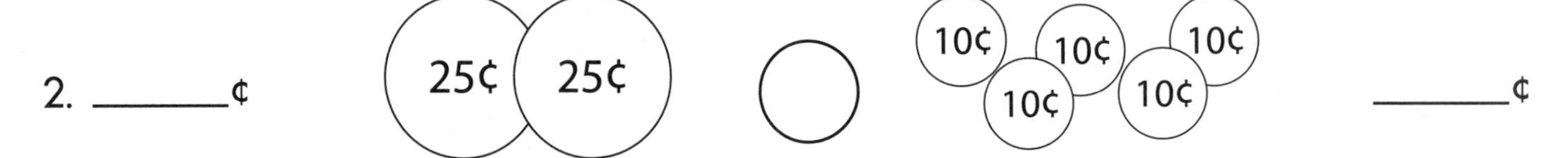

______¢

3. ______¢ 1¢ 1¢ 1¢ 1¢ 1¢ 5¢ 10¢ ◯ 25¢ ______¢

4. ______¢ 25¢ 5¢ ◯ 10¢ 10¢ 5¢ ______¢

5. ______¢ 10¢ 10¢ 10¢ 10¢ 10¢ 10¢ 10¢ 10¢ 10¢ ◯ 25¢ 25¢ 25¢ ______¢

Draw your own coin combinations to make each symbol true.

6. $<$

7. $>$

8. $=$

Name ______________________ 2.NBT.A.3, 2.NBT.A.4

Reading and Writing Numbers

Use the numbers below to answer the questions.

578	341	259	97	820	443

Each number above is written in standard form. Write the standard form of each number on the line in a box. Then, draw the picture form of each number.

1.  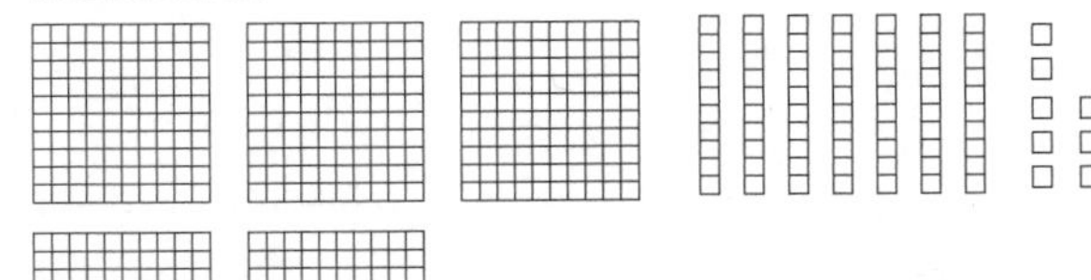	2. ______
3. ______	4. ______
5. ______	6. ______

Use each number at least once to make true comparisons.

7. ______ > ______

8. ______ > ______

9. ______ < ______

10. ______ < ______

11. ______ = ______

12. ______ = ______

Name ______________________________ 2.NBT.A.3

Reading and Writing Numbers

Write the next three numbers in each pattern. Circle the last one.
Write the expanded form and draw the representational form of the number you circled.

Use these symbols to draw each number: □ = 100 | = 10 ▫ = one

1. 976, 975, 974, 973, ______ , ______ , ______

   Expanded form: Representational form:

2. 100, 200, 300, ______ , ______ , ______

   Expanded form: Representational form:

3. 367, 377, 387, 397, ______ , ______ , ______

   Expanded form: Representational form:

4. 802, 702, 602, 502, ______ , ______ , ______

   Expanded form: Representational form:

5. 255, 245, 235, 225, ______ , ______ , ______

   Expanded form: Representational form:

6. 338, 348, 358, 368, ______ , ______ , ______

   Expanded form: Representational form:

7. 104, 204, 304, 404, ______ , ______ , ______

   Expanded form: Representational form:

Name ______________________________ 2.NBT.A.3

Reading and Writing Numbers

Look at the place value blocks. Write the standard, expanded, and word forms of each number.

1.

Standard form: __________ Expanded form: ______________________________

Short word form: ______ hundreds, ______ tens, ______ ones

Word form: ______________________________

2.

Standard form: __________ Expanded form: ______________________________

Short word form: ______ hundreds, ______ tens, ______ ones

Word form: ______________________________

3.

Standard form: __________ Expanded form: ______________________________

Short word form: ______ hundreds, ______ tens, ______ ones

Word form: ______________________________

4.

Standard form: __________ Expanded form: ______________________________

Short word form: ______ hundreds, ______ tens, ______ ones

Word form: ______________________________

5.

Standard form: __________ Expanded form: ______________________________

Short word form: ______ hundreds, ______ tens, ______ ones

Word form: ______________________________

Name ____________________

2.NBT.B.5

Adding and Subtracting within 100

Solve each problem.

1. $77 - 3 =$ ____
2. $17 + 0 =$ ____
3. $44 + 2 =$ ____
4. $76 + 3 =$ ____
5. $86 + 4 =$ ____
6. $42 + 4 =$ ____
7. $27 + 1 =$ ____
8. $55 + 2 =$ ____
9. $77 - 6 =$ ____
10. $47 - 6 =$ ____
11. $24 + 4 =$ ____
12. $78 + 1 =$ ____
13. $95 - 0 =$ ____
14. $85 - 0 =$ ____
15. $75 - 0 =$ ____
16. $72 + 1 =$ ____
17. $62 + 6 =$ ____
18. $57 - 6 =$ ____
19. $58 - 8 =$ ____
20. $36 - 5 =$ ____
21. $61 + 3 =$ ____
22. $43 + 3 =$ ____
23. $83 - 0 =$ ____
24. $89 - 1 =$ ____

Name ______________________

2.NBT.B.5

Adding and Subtracting within 100

Solve each problem.

1. 74 − 30	2. 76 + 22	3. 72 + 23	4. 74 + 12	5. 85 − 61	6. 45 + 24
7. 60 + 34	8. 78 + 21	9. 76 − 26	10. 43 − 23	11. 78 + 21	12. 43 + 53
13. 16 − 12	14. 54 − 32	15. 82 − 42	16. 33 + 33	17. 75 + 24	18. 64 − 23
19. 45 − 21	20. 76 − 25	21. 54 + 45	22. 67 + 22	23. 66 − 51	24. 83 − 62
25. 52 − 31	26. 34 + 22	27. 43 + 34	28. 80 − 20	29. 66 + 10	30. 42 + 42

Name ______________________________ 2.NBT.B.5

Adding and Subtracting within 100

Solve each problem.

1. $27 + 82$	2. $74 + 95$	3. $80 - 60$	4. $95 + 44$	5. $94 - 93$	6. $41 + 75$
7. $82 - 71$	8. $97 - 81$	9. $85 + 91$	10. $33 + 85$	11. $92 + 54$	12. $74 + 80$
13. $11 + 75$	14. $53 + 72$	15. $83 + 84$	16. $52 + 85$	17. $65 - 42$	18. $94 - 64$
19. $73 + 74$	20. $95 - 33$	21. $44 + 92$	22. $52 + 66$	23. $71 + 66$	24. $60 + 85$
25. $93 + 26$	26. $32 + 72$	27. $44 + 84$	28. $81 - 40$	29. $114 - 10$	30. $158 - 13$

Name ______________________________ 2.NBT.B.6

Adding up to Four Two-Digit Numbers with Expanded Notation

Add two two-digit numbers together by using **expanded notation**.
35 + 29 = 64

35 is 30 + 5
29 is 20 + 9
50 + 14 = 64

Solve each problem. Show your work using expanded notation.

1. 27 + 26
   27 is 20 + 7
   26 is 20 + 6
   40 + 13 = 53

2. 49 + 31

3. 53 + 61

4. 65 + 24

5. 77 + 52

6. 82 + 65

Name ______________________________ 2.NBT.B.6

Adding up to Four Two-Digit Numbers with Expanded Notation

Add three two-digit numbers together by using **expanded notation**.
67 + 35 + 29 = 131

67 is 60 + 7

35 is 30 + 5

29 is 20 + 9

110 + 21 = 131

Solve each problem. Show your work using expanded notation.

1.
$$\begin{array}{r} 27 \\ 10 \\ +\ 25 \\ \hline \end{array}$$

2.
$$\begin{array}{r} 33 \\ 12 \\ +\ 39 \\ \hline \end{array}$$

3.
$$\begin{array}{r} 29 \\ 20 \\ +\ 53 \\ \hline \end{array}$$

4.
$$\begin{array}{r} 48 \\ 31 \\ +\ 22 \\ \hline \end{array}$$

5.
$$\begin{array}{r} 72 \\ 53 \\ +\ 11 \\ \hline \end{array}$$

6.
$$\begin{array}{r} 47 \\ 42 \\ +\ 53 \\ \hline \end{array}$$

Name ______________________________ (2.NBT.B.6)

Adding up to Four Two-Digit Numbers with Expanded Notation

Add four two-digit numbers together by using **expanded notation**.
44 + 67 + 35 + 29 = 175

44 is 40 + 4

67 is 60 + 7

35 is 30 + 5

29 is 20 + 9

150 + 25 = 175

Solve each problem. Show your work using expanded notation.

1.
```
  27
  16
  64
+ 18
----
```

2.
```
  12
  25
  36
+ 15
----
```

3.
```
  21
  11
  29
+ 42
----
```

4.
```
  29
  51
  53
+ 30
----
```

5.
```
  50
  24
  35
+ 36
----
```

6.
```
  12
  22
  47
+ 28
----
```

Name ______________________________ 2.NBT.B.7

Adding Three-Digit Numbers within 1,000

Decompose numbers to help you add three-digit numbers.

$$\begin{array}{r} 210 \\ +\ 125 \\ \hline \end{array}$$

$$\begin{array}{rrr} 200 + & 100 = & 300 \\ 10 + & 20 = & 30 \\ 0 + & 5 = & 5 \\ \hline & & 335 \end{array}$$

Solve each problem. Show your work.

1. $\begin{array}{r} 227 \\ +\ 131 \\ \hline \end{array}$

_____ + _____ = _____
_____ + _____ = _____
_____ + _____ = _____
_____

2. $\begin{array}{r} 331 \\ +\ 256 \\ \hline \end{array}$

_____ + _____ = _____
_____ + _____ = _____
_____ + _____ = _____
_____

3. $\begin{array}{r} 516 \\ +\ 142 \\ \hline \end{array}$

4. $\begin{array}{r} 427 \\ +\ 221 \\ \hline \end{array}$

Name ______________________________ 2.NBT.B.7

Adding Three-Digit Numbers within 1,000

Decompose numbers or use other methods, like place value blocks, to help you add three-digit numbers.

$$\begin{array}{r} 261 \\ +\ 227 \\ \hline 200 + 200 = 400 \\ 60 + 20 = 80 \\ 1 + 7 = 8 \\ \hline 488 \end{array}$$

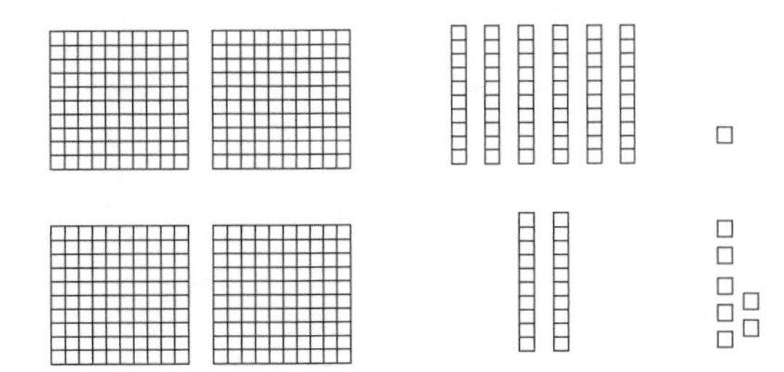

$400 + 80 + 8 = 488$

Solve each problem. Show your work.

1. $\begin{array}{r} 155 \\ +\ 124 \\ \hline \end{array}$

2. $\begin{array}{r} 113 \\ +\ 371 \\ \hline \end{array}$

3. $\begin{array}{r} 375 \\ +\ 313 \\ \hline \end{array}$

4. $\begin{array}{r} 277 \\ +\ 222 \\ \hline \end{array}$

5. $\begin{array}{r} 336 \\ +\ 132 \\ \hline \end{array}$

6. $\begin{array}{r} 482 \\ +\ 211 \\ \hline \end{array}$

Name ______________________________

2.NBT.B.7

Adding Three-Digit Numbers within 1,000

Solve each problem. Show your work.

1. $\begin{array}{r} 486 \\ +\ 313 \\ \hline \end{array}$

2. $\begin{array}{r} 639 \\ +\ 250 \\ \hline \end{array}$

3. $\begin{array}{r} 387 \\ +\ 412 \\ \hline \end{array}$

4. $\begin{array}{r} 563 \\ +\ 416 \\ \hline \end{array}$

5. $\begin{array}{r} 574 \\ +\ 225 \\ \hline \end{array}$

6. $\begin{array}{r} 362 \\ +\ 332 \\ \hline \end{array}$

7. $\begin{array}{r} 667 \\ +\ 300 \\ \hline \end{array}$

8. $\begin{array}{r} 450 \\ +\ 246 \\ \hline \end{array}$

9. $\begin{array}{r} 738 \\ +\ 261 \\ \hline \end{array}$

10. $\begin{array}{r} 113 \\ +\ 215 \\ \hline \end{array}$

Name ____________________ 2.NBT.B.7

Subtracting Three-Digit Numbers within 1,000

Decompose numbers to help you subtract three-digit numbers.

$$\begin{array}{r} 289 \\ -\ 125 \\ \hline \end{array}$$

$$\begin{array}{rcrcr} 200 & - & 100 & = & 100 \\ 80 & - & 20 & = & 60 \\ 9 & - & 5 & = & 4 \\ \hline & & & & 164 \end{array}$$

Solve each problem. Show your work.

1. $\begin{array}{r} 577 \\ -\ 234 \\ \hline \end{array}$

_____ – _____ = _____

_____ – _____ = _____

_____ – _____ = _____

_____

2. $\begin{array}{r} 643 \\ -\ 521 \\ \hline \end{array}$

_____ – _____ = _____

_____ – _____ = _____

_____ – _____ = _____

_____

3. $\begin{array}{r} 498 \\ -\ 257 \\ \hline \end{array}$

4. $\begin{array}{r} 873 \\ -\ 752 \\ \hline \end{array}$

Name ____________________ 2.NBT.B.7

Subtracting Three-Digit Numbers within 1,000

Decompose numbers or use other methods, like place value blocks, to help you subtract three-digit numbers.

```
   465
 − 223
 -----
   400 − 200 = 200
    60 −  20 =  40
     5 −   3 =   2
             -----
               242
```

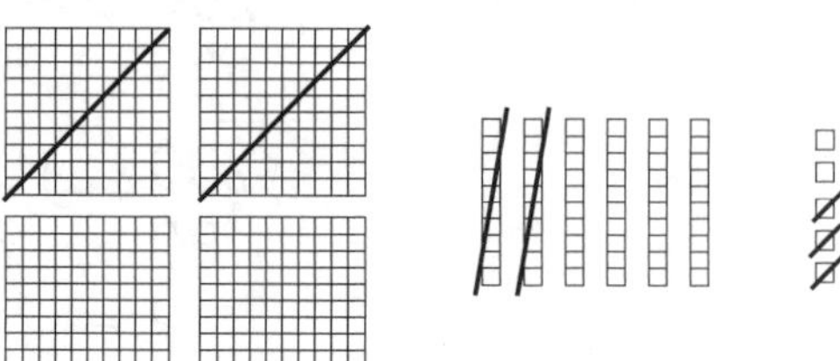

$465 - 223 = 242$

Solve each problem. Show your work.

1. $\begin{array}{r} 653 \\ -\ 412 \\ \hline \end{array}$

2. $\begin{array}{r} 749 \\ -\ 513 \\ \hline \end{array}$

3. $\begin{array}{r} 688 \\ -\ 533 \\ \hline \end{array}$

4. $\begin{array}{r} 495 \\ -\ 253 \\ \hline \end{array}$

5. $\begin{array}{r} 777 \\ -\ 211 \\ \hline \end{array}$

6. $\begin{array}{r} 852 \\ -\ 751 \\ \hline \end{array}$

Name ____________________ 2.NBT.B.7

Subtracting Three-Digit Numbers within 1,000

Solve each problem. Show your work.

1. $\begin{array}{r} 981 \\ -\ 360 \\ \hline \end{array}$

2. $\begin{array}{r} 213 \\ -\ 101 \\ \hline \end{array}$

3. $\begin{array}{r} 999 \\ -\ 222 \\ \hline \end{array}$

4. $\begin{array}{r} 548 \\ -\ 413 \\ \hline \end{array}$

5. $\begin{array}{r} 179 \\ -\ 156 \\ \hline \end{array}$

6. $\begin{array}{r} 887 \\ -\ 544 \\ \hline \end{array}$

7. $\begin{array}{r} 324 \\ -\ 123 \\ \hline \end{array}$

8. $\begin{array}{r} 495 \\ -\ 262 \\ \hline \end{array}$

9. $\begin{array}{r} 474 \\ -\ 333 \\ \hline \end{array}$

10. $\begin{array}{r} 262 \\ -\ 131 \\ \hline \end{array}$

Name ______________________ 2.NBT.B.7

Regrouping

Look at the groups. Each picture shows more than 10 one blocks. They need to be **regrouped**. This means trading 10 of the ones to make another ten.

Example:  = =

3 tens 16 ones 4 tens 6 ones

Make another group of ten and write the number a new way.

1. 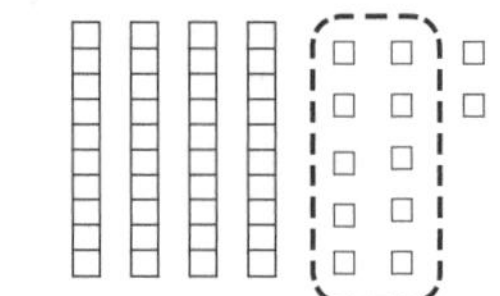

4 tens 12 ones

T	O
5	**2**

2. 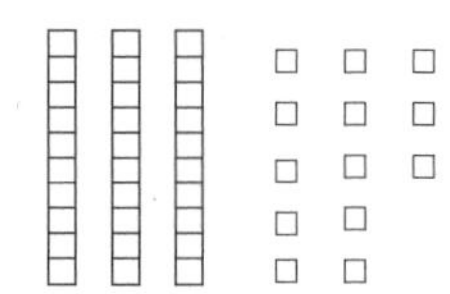

3 tens 13 ones

T	O

3. 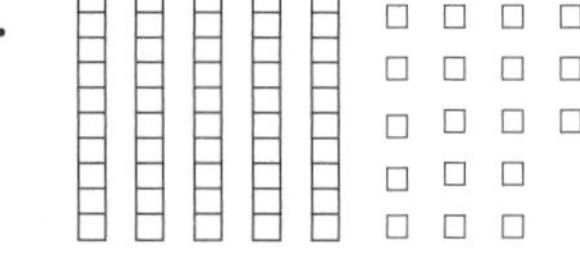

5 tens 18 ones

T	O

4. 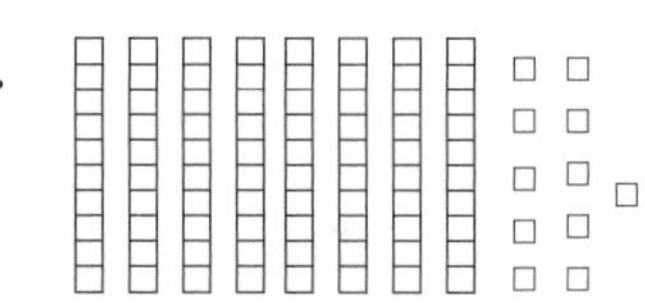

8 tens 11 ones

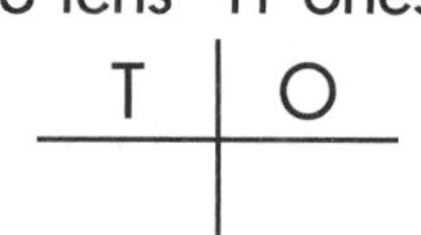

T	O

5. 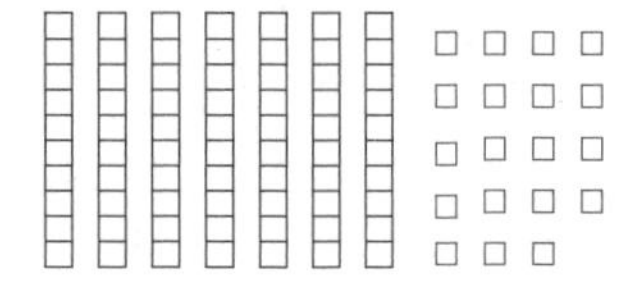

7 tens 19 ones

T	O

6. 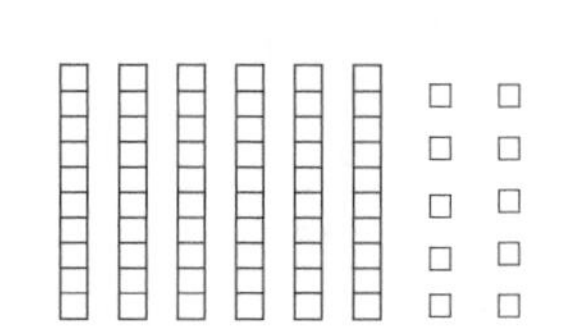

6 tens 10 ones

T	O

Now, try regrouping without pictures. Write the new number in the tens and ones columns.

7. 2 tens 11 ones

T	O

8. 7 tens 10 ones

T	O

9. 1 ten 18 ones

T	O

10. 6 tens 14 ones

T	O

11. 5 tens 19 ones

T	O

12. 8 tens 13 ones

T	O

Name ______________________ 2.NBT.B.7

Regrouping

Write each problem in short word form. Add the ones. Add the tens. Regroup. Write the answer in number form.

1. 72 + 83 can be written as:

   7 tens + 2 ones
   + 8 tens + 3 ones

   15 tens + **5** ones

   Regroup:
   1 hundreds + **5** tens + **5** ones = **155**

2. 54 + 38 can be written as:

   5 tens + 4 ones
   + 3 tens + 8 ones

   ____ tens + ____ ones

   Regroup:
   ____ hundreds + ____ tens + ____ ones = ____

3. 97 + 72 can be written as:

   ____ tens + ____ ones
   + ____ tens + ____ ones

   ____ tens + ____ ones

   Regroup:
   ____ hundreds + ____ tens + ____ ones = ____

4. 57 + 28 can be written as:

   ____ tens + ____ ones
   + ____ tens + ____ ones

   ____ tens + ____ ones

   Regroup:
   ____ hundreds + ____ tens + ____ ones = ____

5. 68 + 71 can be written as:

   ____ tens + ____ ones
   + ____ tens + ____ ones

   ____ tens + ____ ones

   Regroup:
   ____ hundreds + ____ tens + ____ ones = ____

6. 61 + 29 can be written as:

   ____ tens + ____ ones
   + ____ tens + ____ ones

   ____ tens + ____ ones

   Regroup:
   ____ hundreds + ____ tens + ____ ones = ____

Name ______________________________ 2.NBT.B.7

Regrouping

Rewrite each problem. Add and then regroup.

1. 39 + 42 = ______

2. 73 + 18 = ______

3. 572 + 119 = ______

4. 325 + 129 = ______

5. 756 + 104 = ______

6. 667 + 124 = ______

7. 357 + 328 = ______

8. 162 + 539 = ______

Name ______________________________ 2.NBT.B.8

Adding or Subtracting 10

Subtracting 10 from any number only changes the tens place.

Take 10 seeds away from each picture. Write the number.

1.    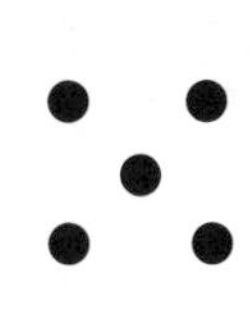

$35 - 10 =$ **25**

2. 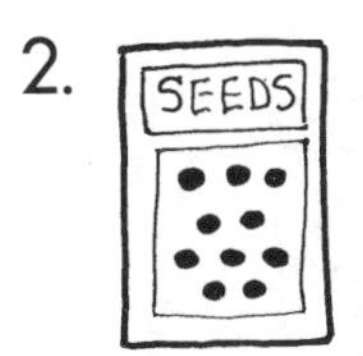     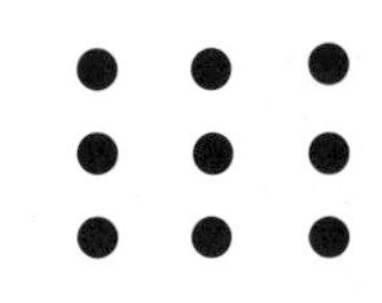

$59 - 10 =$ _____

3. 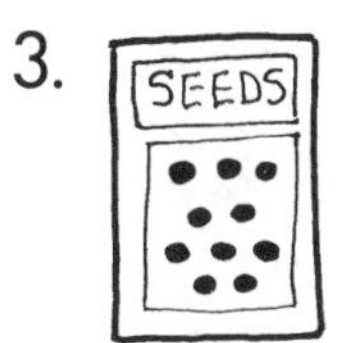 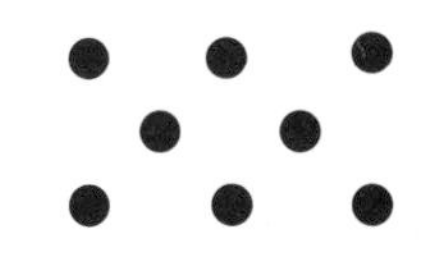

$18 - 10 =$ _____

4. 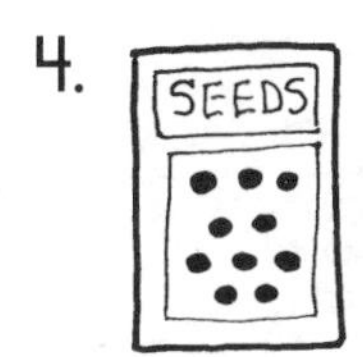       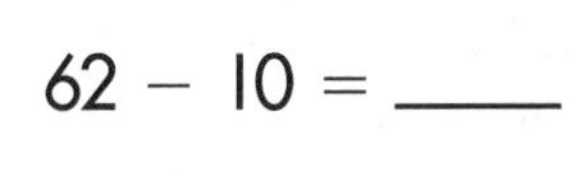

$62 - 10 =$ _____

5. 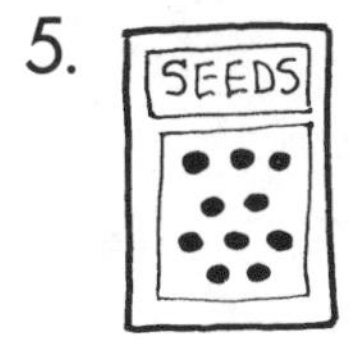      

$70 - 10 =$ _____

6.     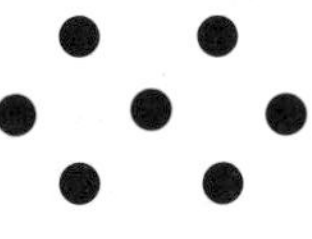

$47 - 10 =$ _____

Name ______________________________ (2.NBT.B.8)

Adding or Subtracting 10

Take 10 away from each picture. Write the new number.

1. 33 − 10 = **23**

2. 46 − 10 = ____

3. 75 − 10 = ____

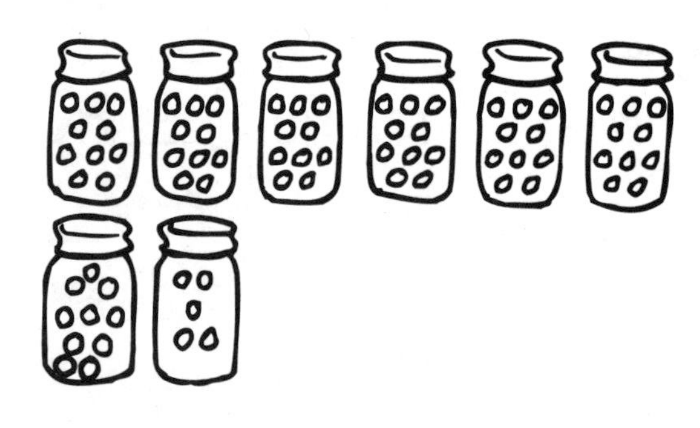

4. 68 − 10 = ____

5. 90 − 10 = ____

6. 58 − 10 = ____

7. Complete the tables by subtracting 10 from each number on the left.

− 10	
59	**49**
46	
12	
74	

− 10	
65	
28	
83	
37	

Name ______________________ 2.NBT.B.8

Adding or Subtracting 10

Look at each number in the middle column. Write the number that is 10 less and the number that is 10 more.

	10 Less		10 More		10 Less		10 More
1.	______	16	______	2.	______	18	______
3.	______	17	______	4.	______	19	______
5.	______	11	______	6.	______	15	______
7.	______	12	______	8.	______	14	______
9.	______	19	______	10.	______	22	______
11.	______	25	______	12.	______	28	______
13.	______	30	______	14.	______	36	______
15.	______	41	______	16.	______	45	______
17.	______	53	______	18.	______	67	______
19.	______	88	______	20.	______	94	______

Name ______________________________

Adding and Subtracting 100s

Adding and subtracting groups of 100s only changes the hundreds place.

Scientists are always looking at our changing star patterns. Imagine hundreds of new stars being discovered while hundreds of others burn out.

Complete the star chart by adding and subtracting 100.

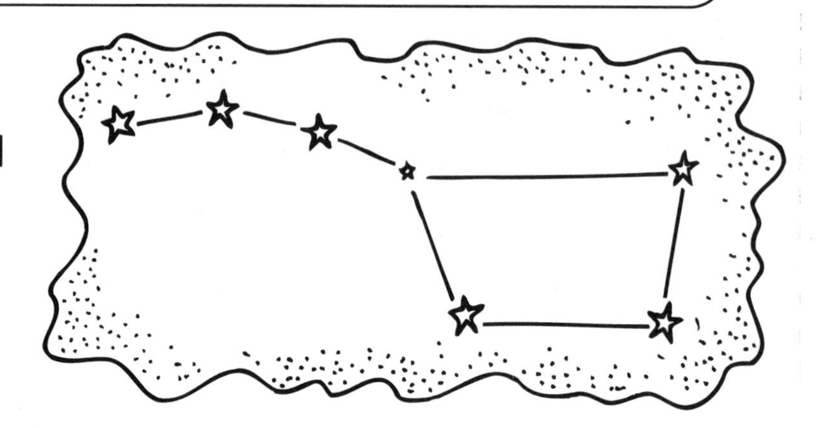

	Number of ★ Before	+ Discovered ★	= Number of ★ Now
1.	881	+ 100	**981**
2.	763	+ 200	
3.	405	+ 300	
4.	312	+ 400	

	Number of ★ Before	– Burned Out ★	= Number of ★ Now
5.	962	– 200	**762**
6.	814	– 300	
7.	730	– 400	
8.	689	– 500	
9.	978	– 600	

10. Explain how you know 300 – 100 = 200. ______________________________

____

Name ______________________________ 2.NBT.B.8

Adding and Subtracting 100s

Look at each number in the middle column. Write the number that is 100 less and the number that is 100 more.

	100 Less		100 More		100 Less		100 More
1.	______	206	______	2.	______	799	______
3.	______	514	______	4.	______	491	______
5.	______	307	______	6.	______	844	______
7.	______	129	______	8.	______	471	______
9.	______	662	______	10.	______	222	______
11.	______	225	______	12.	______	628	______
13.	______	330	______	14.	______	363	______
15.	______	410	______	16.	______	549	______
17.	______	513	______	18.	______	670	______
19.	______	808	______	20.	______	900	______

Name ____________________ 2.NBT.B.8

Adding and Subtracting 100s

Look for key words that tell you to add or subtract. Solve.

1. Fran read 100 pages on Saturday. She read 87 pages on Sunday. How many more pages did Fran read on Saturday?	2. Carlos counted 200 books in his classroom. Troy counted 88 books. How many books did they count in all?
3. Jasmine read for 63 minutes in the morning. After lunch, she read for 100 minutes. Before she went to bed, she read for 10 minutes. How many minutes did Jasmine read altogether?	4. Molly bought a book with 315 pages. She has read 215 pages. How many pages does she have left?
5. Emma counted 397 words in her book. Andy counted 197 words in his book. How many more words are in Emma's book?	6. Grant has read 200 books from the library. Jason has read 125 books. How many books have they read in all?

Name ______________________________ 2.NBT.B.8

Addition and Subtraction with 10 or 100

Solve each problem.

1. $\begin{array}{r} 26 \\ -\ 10 \\ \hline \end{array}$
2. $\begin{array}{r} 37 \\ +\ 10 \\ \hline \end{array}$
3. $\begin{array}{r} 98 \\ -\ 10 \\ \hline \end{array}$
4. $\begin{array}{r} 94 \\ -\ 10 \\ \hline \end{array}$
5. $\begin{array}{r} 66 \\ +\ 10 \\ \hline \end{array}$
6. $\begin{array}{r} 76 \\ -\ 10 \\ \hline \end{array}$
7. $\begin{array}{r} 34 \\ +\ 10 \\ \hline \end{array}$
8. $\begin{array}{r} 52 \\ +\ 10 \\ \hline \end{array}$
9. $\begin{array}{r} 31 \\ +\ 10 \\ \hline \end{array}$
10. $\begin{array}{r} 80 \\ +\ 10 \\ \hline \end{array}$
11. $\begin{array}{r} 86 \\ +\ 10 \\ \hline \end{array}$
12. $\begin{array}{r} 37 \\ +\ 10 \\ \hline \end{array}$
13. $\begin{array}{r} 63 \\ +\ 10 \\ \hline \end{array}$
14. $\begin{array}{r} 80 \\ -\ 10 \\ \hline \end{array}$
15. $\begin{array}{r} 97 \\ -\ 10 \\ \hline \end{array}$
16. $\begin{array}{r} 93 \\ -\ 10 \\ \hline \end{array}$
17. $\begin{array}{r} 83 \\ +\ 10 \\ \hline \end{array}$
18. $\begin{array}{r} 71 \\ -\ 10 \\ \hline \end{array}$

Name ______________________ (2.NBT.B.8)

Addition and Subtraction with 10 or 100

Solve each problem.

1. $81 + 10$
2. $10 + 60$
3. $10 + 43$
4. $63 - 10$
5. $44 - 10$
6. $72 + 10$
7. $61 - 10$
8. $44 + 10$
9. $10 + 42$
10. $19 - 10$
11. $37 - 10$
12. $10 + 32$
13. $10 + 75$
14. $62 - 10$
15. $10 + 32$
16. $10 + 21$
17. $21 - 10$
18. $10 + 41$
19. $10 + 53$
20. $42 - 10$
21. $60 - 10$
22. $10 + 46$
23. $10 + 61$
24. $111 - 10$

Name ______________________________ 2.NBT.B.8

Addition and Subtraction with 10 or 100

Solve each problem.

1. $651 - 10$
2. $994 - 10$
3. $896 - 10$
4. $862 - 10$
5. $379 - 10$
6. $977 - 10$
7. $795 - 100$
8. $785 - 100$
9. $996 - 100$
10. $899 - 100$
11. $439 - 100$
12. $769 - 100$
13. $983 + 10$
14. $968 + 10$
15. $543 + 10$
16. $978 + 10$
17. $398 + 10$
18. $300 + 10$
19. $825 + 100$
20. $170 + 100$
21. $286 + 100$
22. $487 + 100$
23. $664 - 100$
24. $856 + 100$

25. Explain how to add 10 to a number. ______________________________

____

____

Name ______________________ 2.NBT.A.3, 2.NBT.A.4, 2.NBT.B.9

Explain Your Reasoning

Explain your reasoning when comparing numbers by looking at the place values of the two numbers.

five tens and five ones > 32

This is true because there are 5 tens in 55 and only 3 tens in 32.

60 + 7 < four tens and three ones

This is false because there are more tens in 67 than in 43.

45 = forty-five

This is true because there are the same amount of tens and ones in each number.

Are these comparisons true or false? Circle **True** or **False**. Explain your reasoning.

1. twenty-nine > 30 + 4 True False

   Why? ______________________

2. sixty = 40 + 10 True False

   Why? ______________________

3. 53 > eight tens and two ones True False

   Why? ______________________

4. six tens + two ones < 80 + 9 True False

   Why? ______________________

5. forty-two = 42 True False

   Why? ______________________

Name ____________________ (2.NBT.A.3, 2.NBT.A.4, 2.NBT.B.9)

Explain Your Reasoning

Explain your reasoning when comparing numbers by looking at the place values of the two numbers.

three hundreds + five tens + five ones > 232

This is a true statement because there are 3 hundreds in 355 and only 2 hundreds in 232.

400 + 60 + 7 > 500 + 40 + 3

This is a false statement because there are more hundreds in 543 than in 467.

145 = one hundred forty-five

This is true because there are the same amount of hundreds, tens, and ones in each number.

Are these comparisons true or false? Circle **True** or **False**. Explain your reasoning.

1. 4 hundreds + 4 ones > 9 tens and 6 ones — True — False

   Why? ______________________________

2. 2 hundreds + 4 tens + 2 ones < 422 — True — False

   Why? ______________________________

3. 555 > 6 hundreds — True — False

   Why? ______________________________

4. 700 + 50 + 8 = Seven hundred fifty-eight — True — False

   Why? ______________________________

5. 3 hundreds + 7 ones > 370 — True — False

   Why? ______________________________

Name ______________________________ 2.NBT.A.3, 2.NBT.A.4, 2.NBT.B.9

Explain Your Reasoning

Are these comparisons true or false? Circle **True** or **False**. Explain your reasoning.

1. 4 tens + 2 hundreds + 2 ones < 422 True False

   Why? ______________________________

2. 988 = Nine hundred ninety-eight True False

   Why? ______________________________

3. 500 + 30 + 3 < 678 True False

   Why? ______________________________

4. Three hundred + ninety + six > 396 True False

   Why? ______________________________

5. 9 ones + six hundreds + twenty > 200 + 60 + 9 True False

   Why? ______________________________

6. four hundred three < 430 True False

   Why? ______________________________

7. 70 + 700 + 2 > seven hundred eighty-seven True False

   Why? ______________________________

8. 561 > 8 hundreds + 6 ones + 7 tens True False

   Why? ______________________________

Name ______________________________ (2.MD.A.1, 2.MD.A.3, 2.MD.A.4)

Measuring in Inches and Centimeters

Have you ever noticed that some rulers have numbers on both sides? One side shows **inches** and the other shows **centimeters**. To measure the length of a line, place the starting point of your ruler on the beginning of the line. Then, read the nearest number at the end of the line.

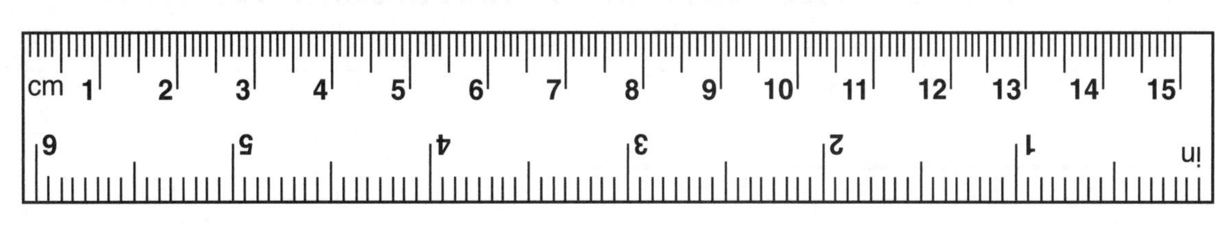

Measure these lines in inches and centimeters. Record the length.

1. ____________________ This line is ______ **inches** long.

2. ____________ This line is ______ **centimeters** long.

3. ____

   This line is ______ **inches** long.

4. ____________________ This line is ______ **centimeters** long.

5. ____

   This line is ______ **inches** long.

6. ____ This line is ______ **inches** long.

7. ____

   This line is ______ **centimeters** long.

8. ____________________________________ This line is ______ **centimeters** long.

9. How much longer is line 5 than 6? ______ inches

10. How much longer is line 7 than 8? ______ centimeters

Name ______________________________ 2.MD.A.1, 2.MD.A.3, 2.MD.A.4

Measuring in Inches and Centimeters

Estimate the length of each line in inches or centimeters. Measure the lines. Record the actual measurement. Write each length.

	Estimate	Actual
1. ____________________________________	_______ in.	_______ in.
2. ______________________	_______ cm	_______ cm
3. _________	_______ in.	_______ in.
4. ___________________________	_______ cm	_______ cm
5. _____	_______ cm	_______ cm
6. _______________	_______ in.	_______ in.
7. ____	_______ cm	_______ cm

Draw a line to match the given measurement.

8. 6 inches

9. 17 centimeters

10. 8 inches

Name ______________________________ (2.MD.A.1, 2.MD.A.3, 2.MD.A.4)

Measuring in Inches and Centimeters

Estimate the length of each line in inches or centimeters. Measure the lines. Record the actual measurement. Write each length.

	Estimate	Actual
1. ____________________	______ cm	______ cm
2. ____	______ in.	______ in.
3. __________	______ cm	______ cm
4. ____	______ in.	______ in.
5. ______________________________	______ in.	______ in.

Draw a line to match the given measurement.

6. 16 centimeters

7. 9 centimeters

8. 7 inches

9. 3 inches

10. How much longer is line 6 than line 7? ______________________________

11. How much longer is line 8 than line 9? ______________________________

Name ______________________ 2.MD.B.5, 2.MD.B.6

Relating Addition and Subtraction to Length

A number line is like a ruler. You can use a number line to add or subtract lengths in a word problem.

Mia used 3 inches of blue yarn and 4 inches of red yarn to make a bracelet for her doll. How many inches of yarn did Mia use all together?

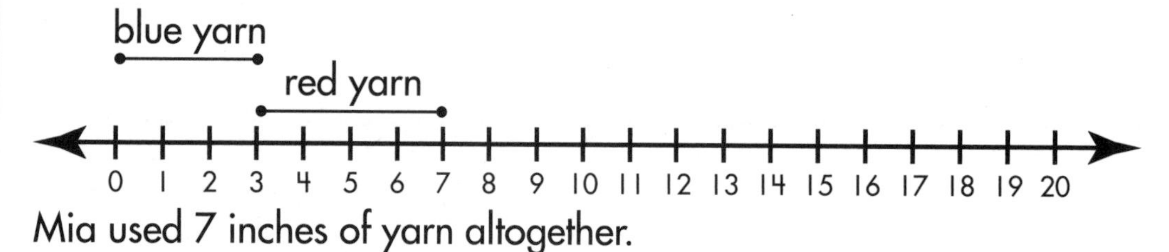

Mia used 7 inches of yarn altogether.

Jake had 12 yards of rope. He gave his friend 7 yards. How many yards of rope does Jake have now?

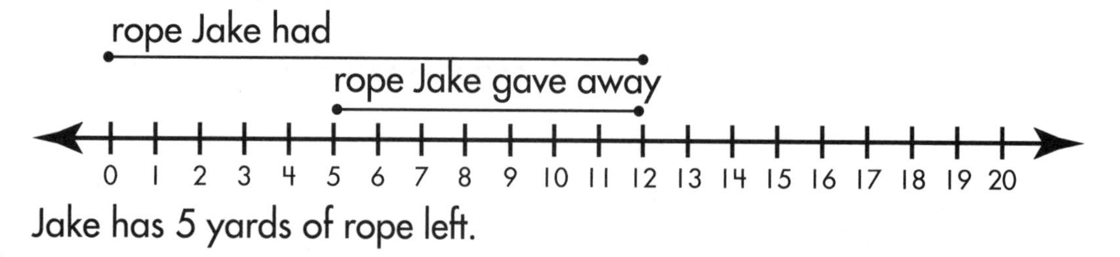

Jake has 5 yards of rope left.

Solve each word problem using the number line.

1. Kelly had 17 feet of ribbon. She gave Chris 6 feet. How much ribbon does she have left?

_____ feet

2. Paul had 18 meters of fishing line. Then, 9 meters broke off. How many meters are left?

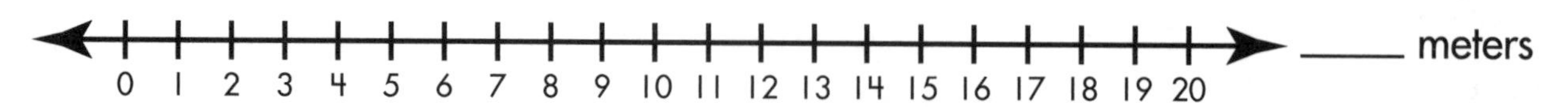

_____ meters

3. Dante has 8 yards of kite string. He needs 12 more yards. How many yards does he need altogether?

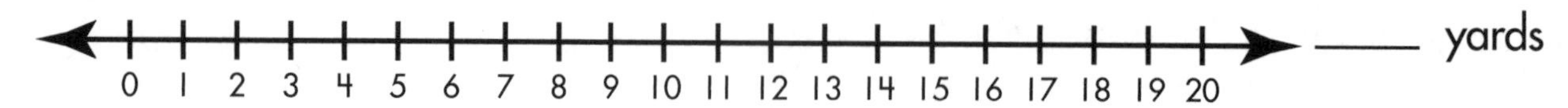

_____ yards

Name ______________________________ 2.MD.B.5, 2.MD.B.6

Relating Addition and Subtraction to Length

Draw a number line to solve each problem.

1. Cole had 27 feet of wire. He gave Dante 8 feet. How much wire does Cole have left?

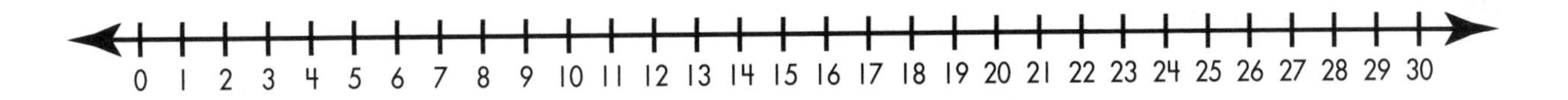

_______ feet

2. Emily had 3 yards of string to fly a kite. The string broke and Emily had only 20 yards of string left. How many yards of string broke?

_______ yards

3. Monica has a 14-inch piece of trim to put on a dress. She needs 12 more inches of trim to finish the dress. How many inches does she need altogether?

_______ inches

4. Before Owen sharpened his pencil, it was 16 centimeters long. After he sharpened it, it was 14 centimeters. How many centimeters longer was the pencil before he sharpened it?

_______ centimeters

5. Olivia walked 27 meters on Saturday. Mason walked 15 more meters than Olivia on Sunday. How many meters did Mason walk on Sunday?

_______ meters

Name ______________________________ 2.MD.B.5, 2.MD.B.6

Relating Addition and Subtraction to Length

Solve each word problem. Show your work with number lines, number sentences, pictures, or words.

1. Laura had 5 inches of her hair cut. Now, her hair is 14 inches long. How long was Laura's hair before the haircut?

2. Kenneth bought a new 25-foot hose. It is 10 feet longer than his old hose. How many feet long was Kenneth's old hose?

3. Grace needs 42 yards of yarn to make a scarf. She has already used 24 yards. How many more yards does Grace need to finish her scarf?

4. Libby is 12 inches taller than her friend. Her friend is 45 inches tall. How tall is Libby?

5. Jane needs 62 feet of rope to make a swing. She has 42 feet of rope. How much more rope does Jane need to make a swing?

6. Neil saw a 250-foot tall tree in the rain forest. Another tree was 125 feet tall. How many more feet tall is the taller tree?

Name ____________________ 2.MD.C.7

Time to the Nearest Five Minutes

There are two ways to show time.

1:15

Digital clocks show time using numbers.

Analog clocks use hands to show us the hours and minutes.

The minute hand on a clock is the long hand. It takes 5 minutes to move from one number on the clock to the next. Therefore, we can count by fives as the minute hand moves from number to number. To read this clock, we say...

20 minutes past 3:00

3:20

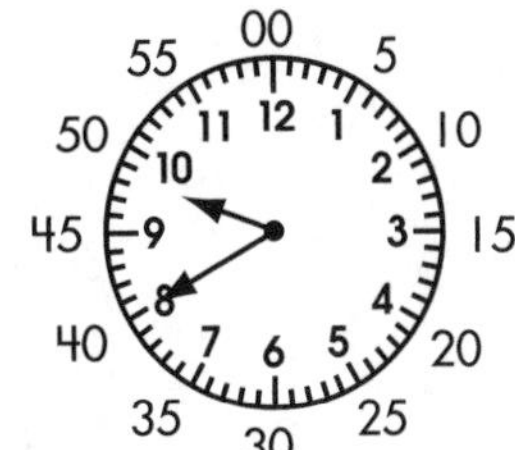

40 minutes past 9:00

9:40

Color the clocks that have matching digital and analog time.

1. 

1:15

2. 

3:25

3. 

7:45

4. 

9:50

5. 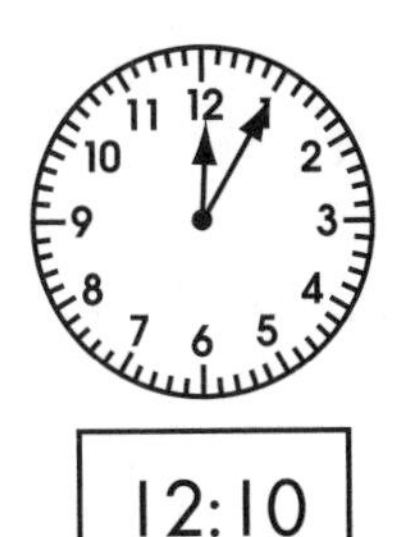

12:10

6. 

8:35

7. 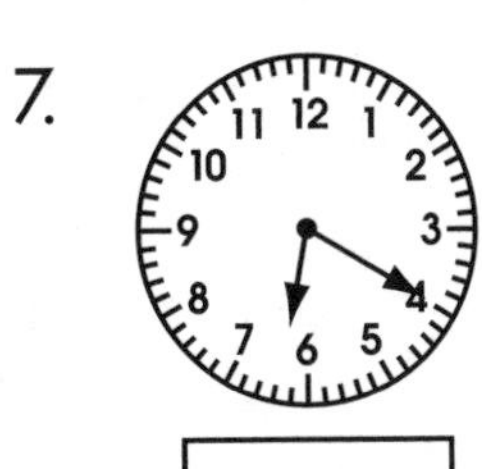

6:20

8. 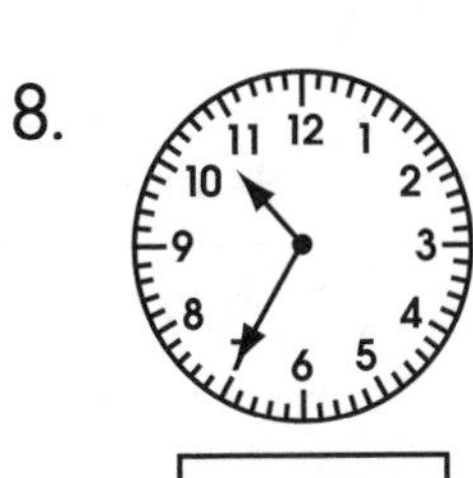

10:30

Name ____________________ 2.MD.C.7

Time to the Nearest Five Minutes

Write the correct time.

1.

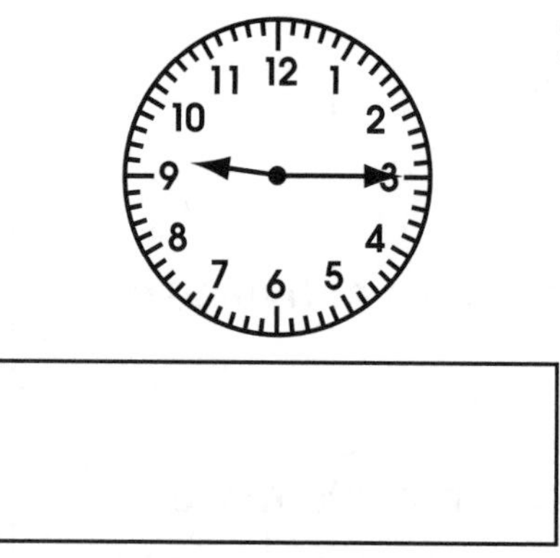

2.

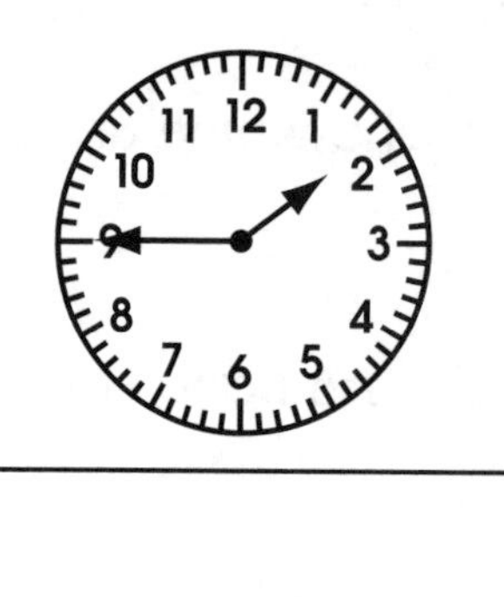

3.

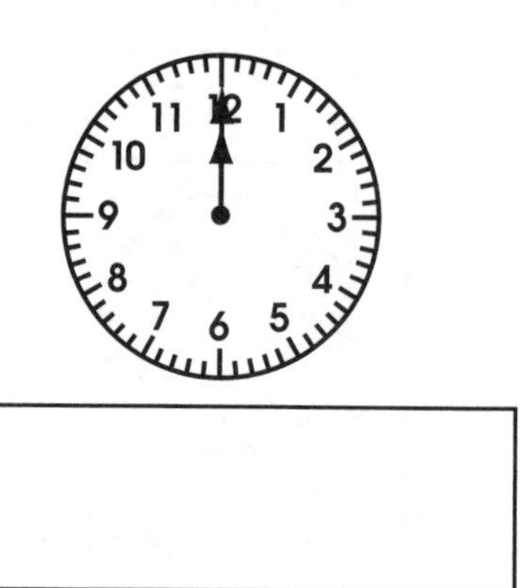

Draw the hands on each clock to show the given time.

4.

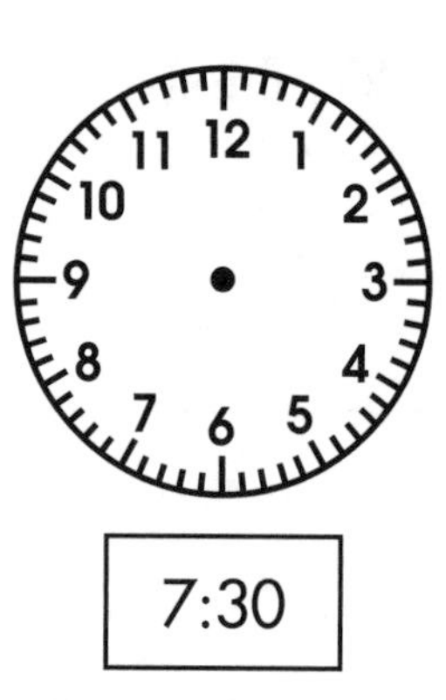

7:30

5.

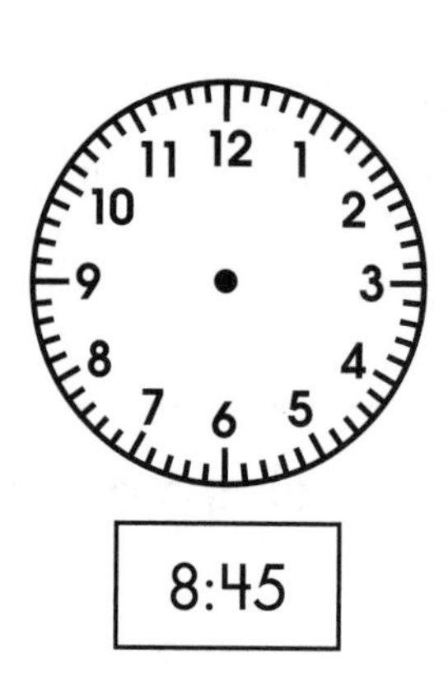

8:45

6.

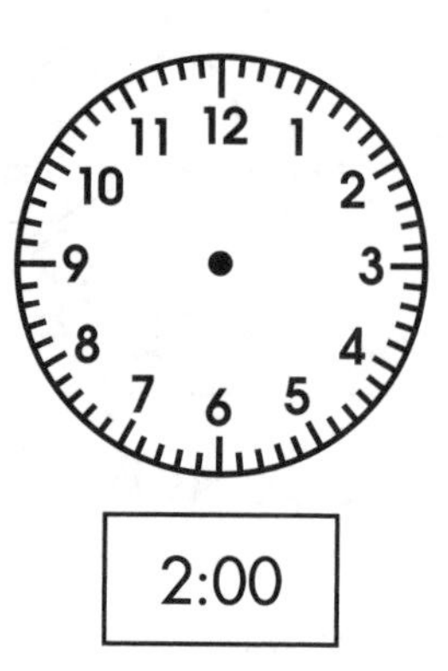

2:00

7.

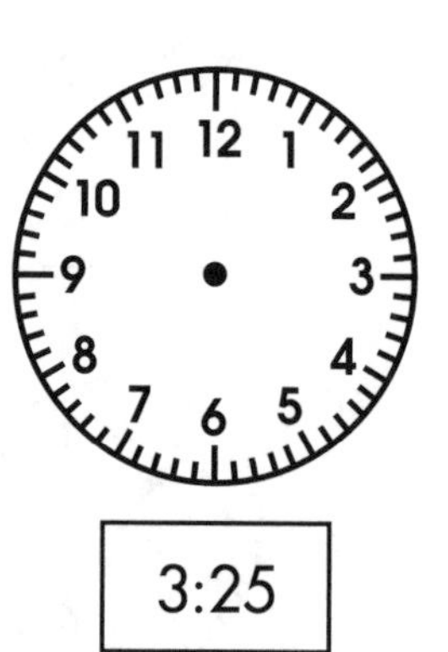

3:25

8.

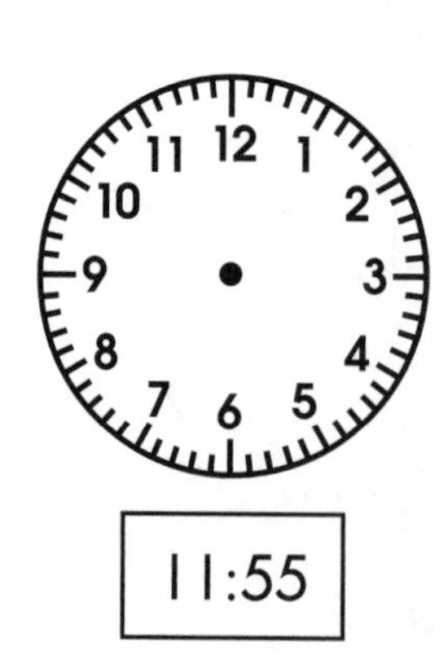

11:55

9.

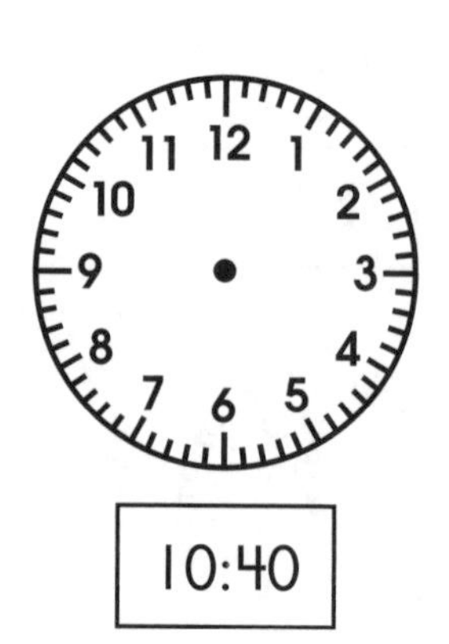

10:40

Name ______________________________

2.MD.C.7

Time to the Nearest Five Minutes

Write the time two ways.

1. [25] minutes past [10:00]
   [10:25]

2. 
   [ ] minutes past [:]
   [:]

3. 
   [ ] minutes past [:]
   [:]

4. 
   [ ] minutes past [:]
   [:]

5. 
   [ ] minutes past [:]
   [:]

6. 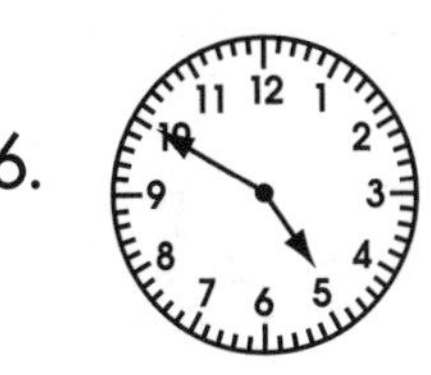
   [ ] minutes past [:]
   [:]

7. 
   [ ] minutes past [:]
   [:]

8. 
   [ ] minutes past [:]
   [:]

9. 
   [ ] minutes past [:]
   [:]

10. 
   [ ] minutes past [:]
   [:]

Name ________________________________ 2.MD.C.8

Counting Money

Here are commonly used coins.

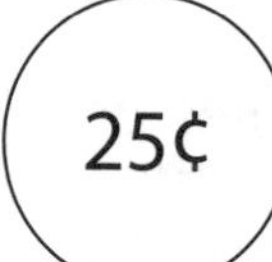 quarter  dime 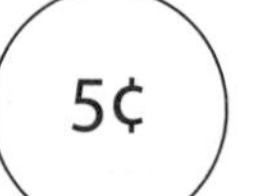 nickel 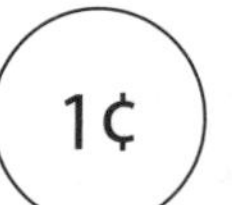 penny

The amount a coin is worth is called its **value**. You can find the total value of coins by adding one amount to another amount, which is called **counting on**.

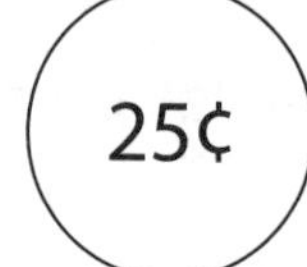 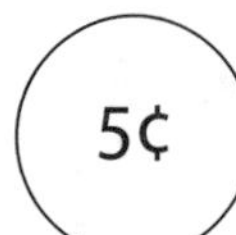 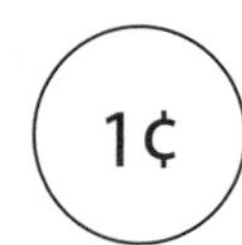

25¢ 35¢ 40¢ 45¢ 46¢ 47¢

Write the value of the coins beside each tree.

1. 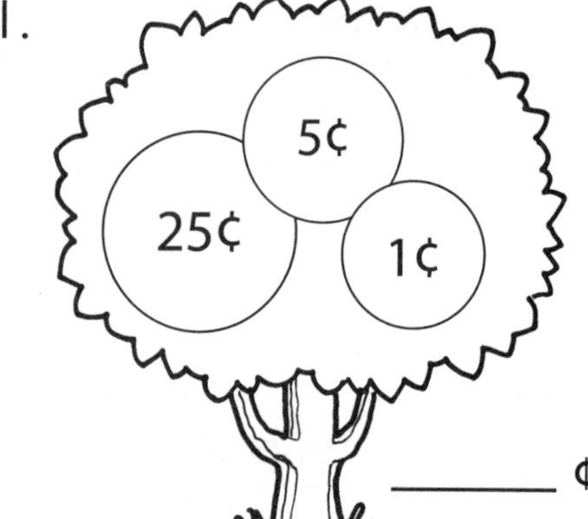

______ ¢

2. 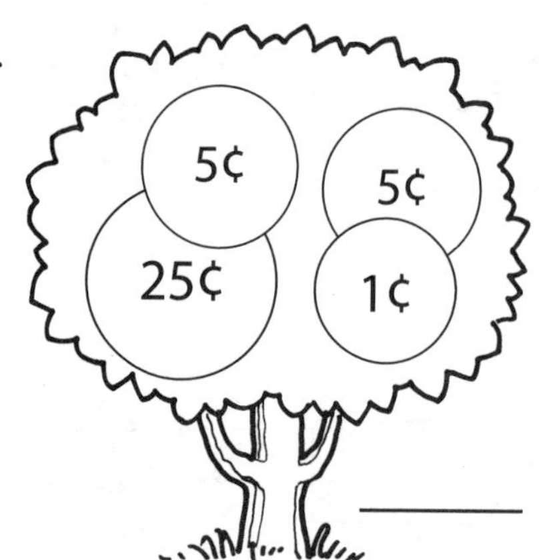

______ ¢

3. 

______ ¢

4. 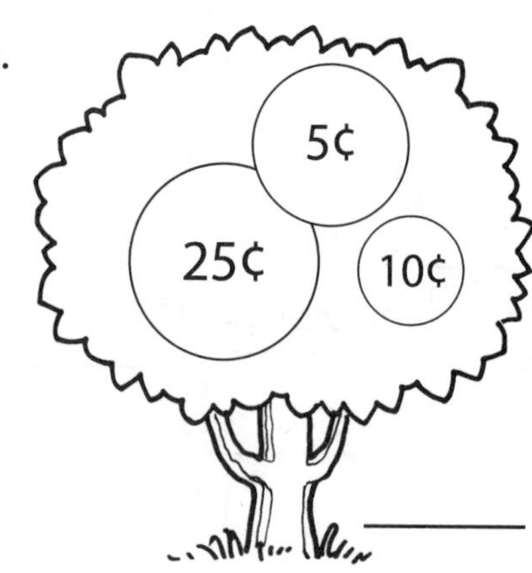

______ ¢

5. 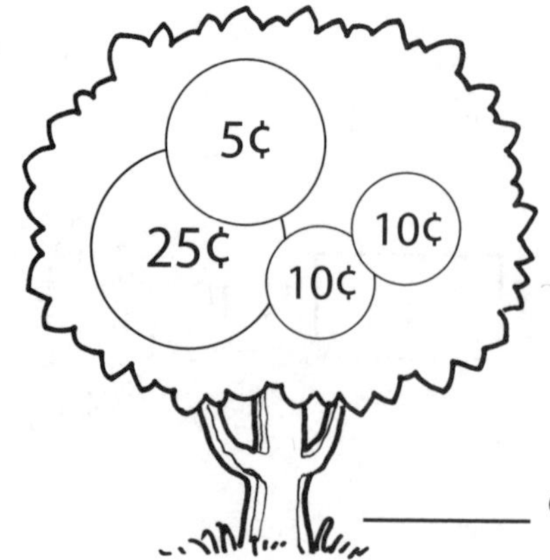

______ ¢

6. 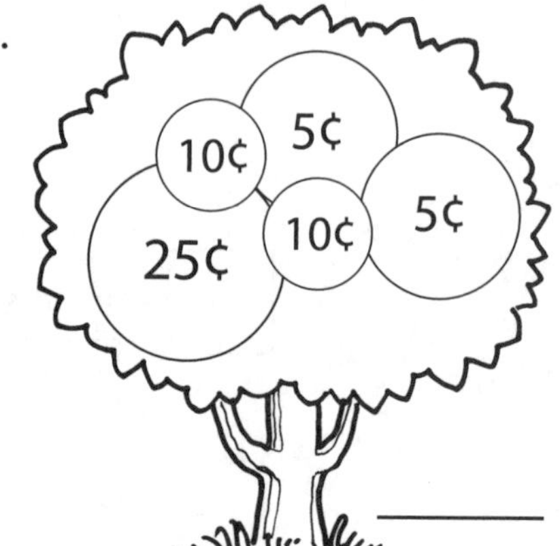

______ ¢

Name ______________________________ 2.MD.C.8

Counting Money

The best way to count money is to begin with the most valuable coins and work your way down.

Example:

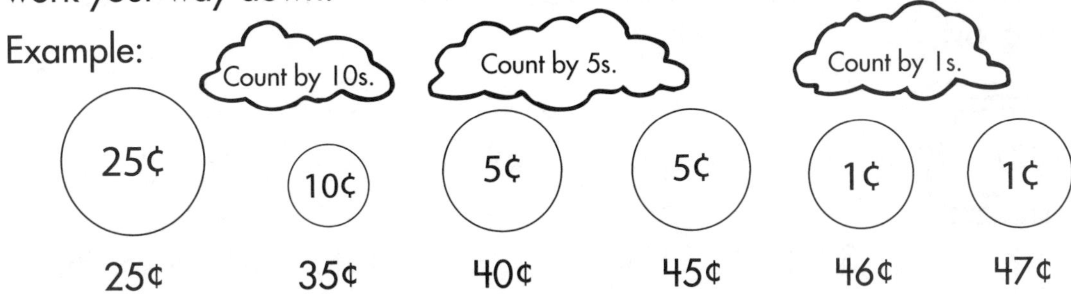

Write the value of each tree's coins on its trunk.

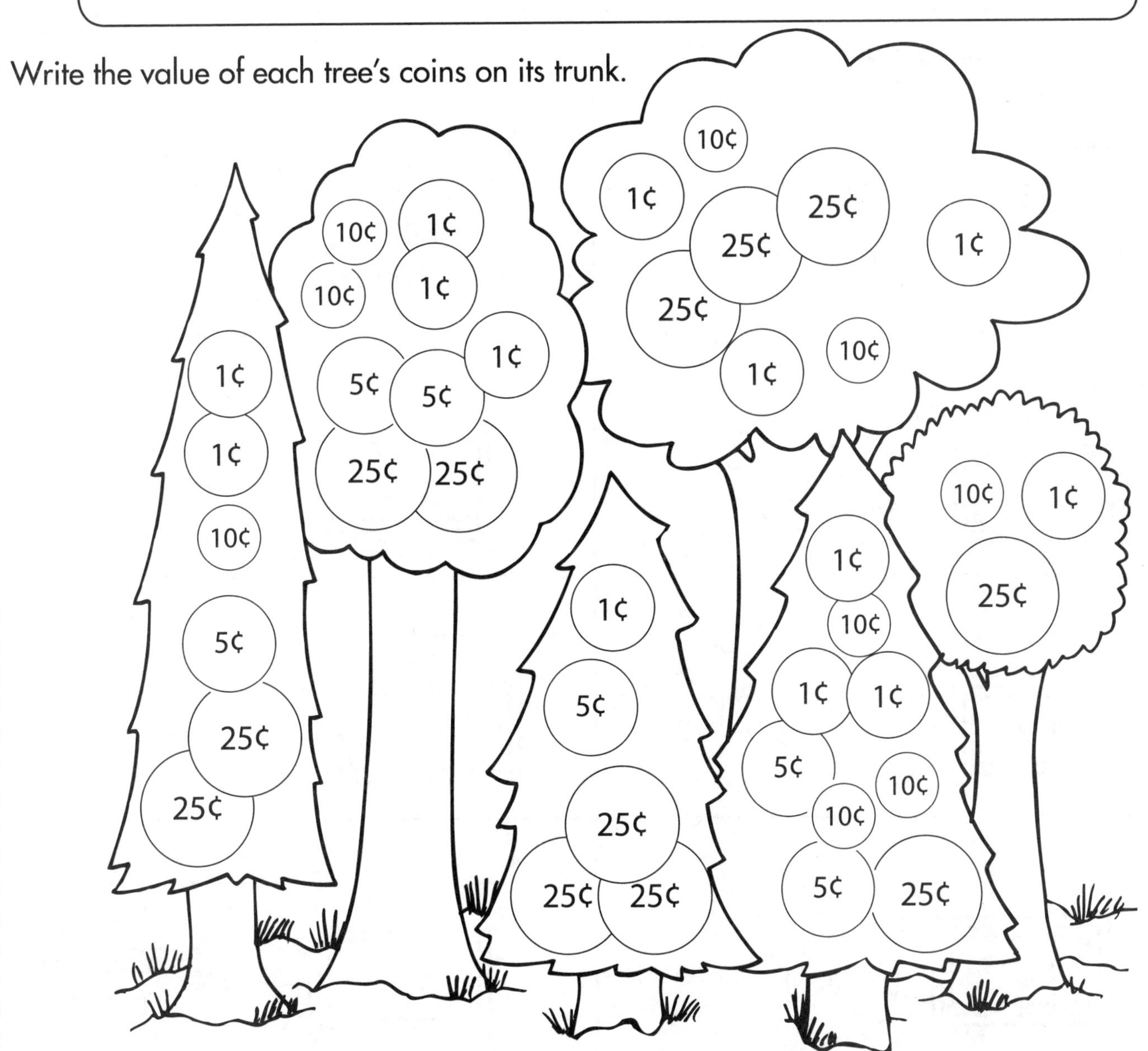

Name ______________________ 2.MD.C.8

Counting Money

The amount a coin is worth is called its **value**. You can find the total value of coins by adding one amount to another amount, which is called **counting on**.

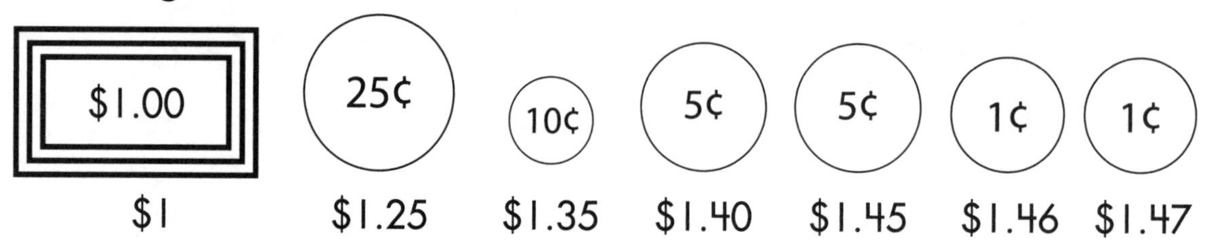

Count on to find the total value.

1. $ ______

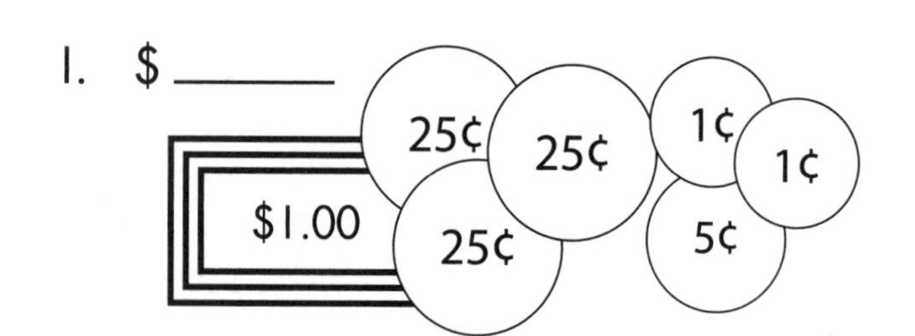

2. $ ______

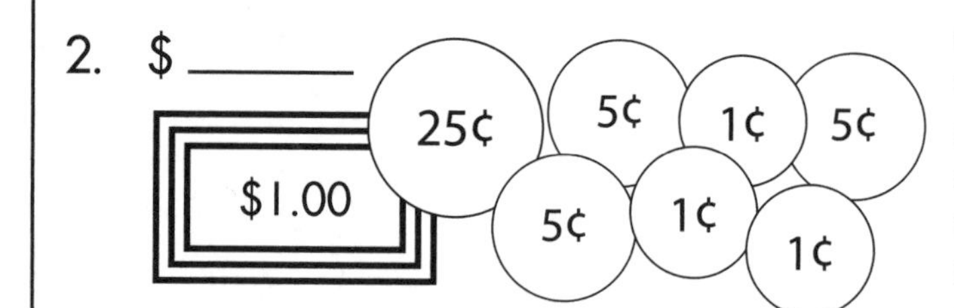

3. $ ______

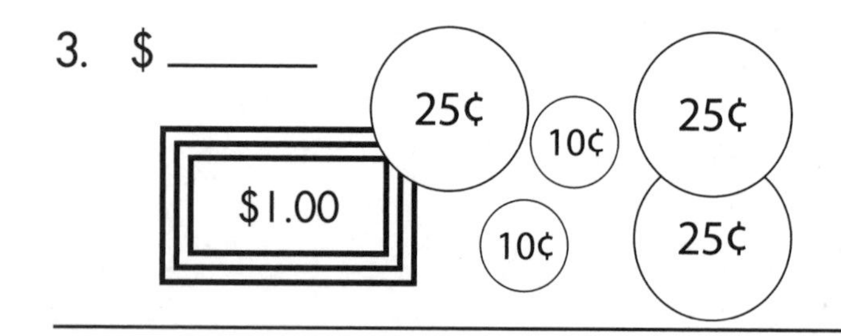

4. $ ______

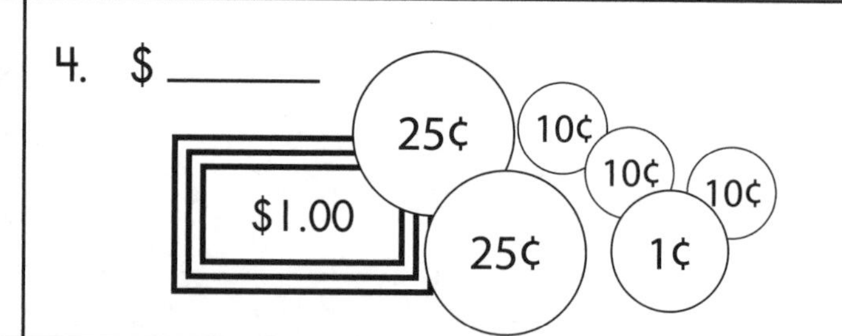

5. $ ______

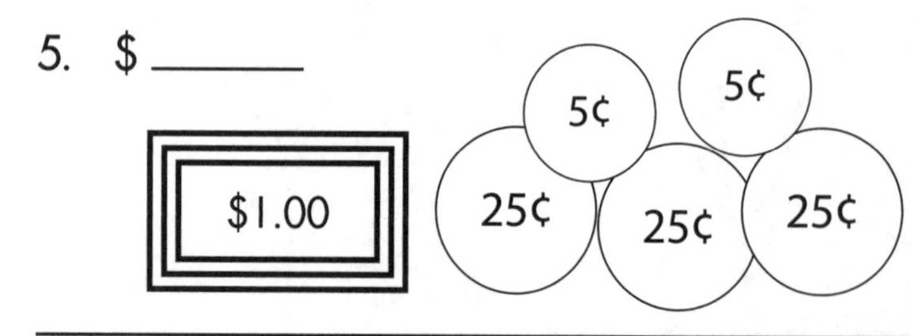

6. $ ______

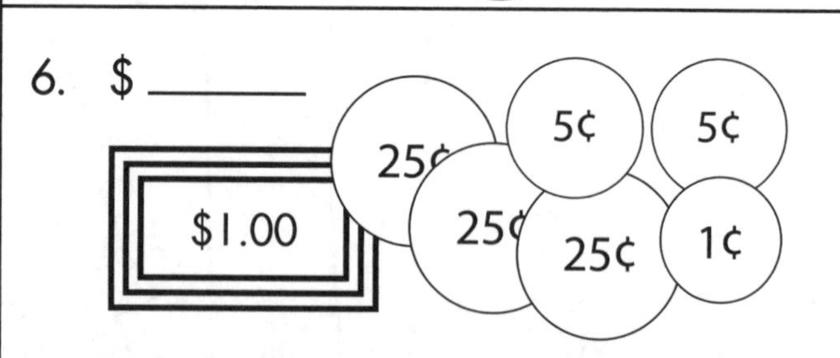

7. $ ______

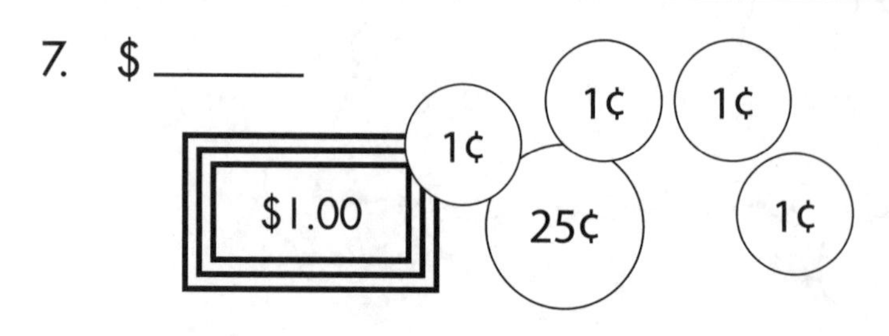

8. $ ______

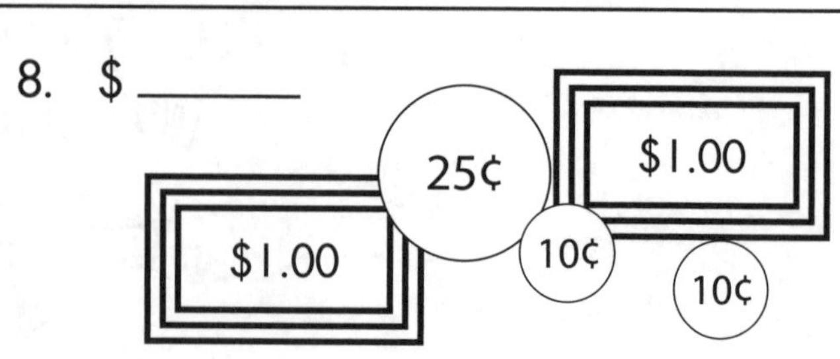

Name ______________________ 2.MD.C.8

Money Problem Solving

Use the information from the story to solve each math problem. Read each story carefully and decide if you should add or subtract to find the answers.

Add or subtract to find the answers.

1. Nell has 9 pennies. She gives 1 to Judy and 3 to Rick.

   How many pennies does Nell have left? __________

   What is their value? __________

2. Hailey has 12 dimes. She gives 5 to Blake and 3 to Jo. Dan gives her 4 more.

   How many dimes does Hailey have now? __________

   What is their value? __________

3. David has 5 nickels. He finds 6 more.

   How many nickels does David have now? __________

   What is their value? __________

4. Jessie has 16 pennies. She gives Jeni 9 pennies and then finds 2 more.

   How many pennies does Jessie have in all? __________

   What is their value? __________

5. Sydney has 8 nickels. Jordi has 5 nickels.

   How many more nickels does Sydney have than Jordi? __________

   What is their value? __________

6. Hector has 13 dimes. Meg has 9 dimes.

   How many more dimes does Hector have than Meg? __________

   What is their value? __________

Name ____________________ 2.MD.C.8

Money Problem Solving

Read. Draw the coins. Show more than one way if you can.

Make 40¢.	Make 85¢.	Make 53¢.
1. Taka has 1 quarter. What could the other coins be?	2. Fita has 3 quarters. What could the other coins be?	3. Lara has 6 nickels. What could the other coins be?
4. Kasha has 3 dimes. What could the other coins be?	5. Reba has 4 dimes. What could the other coins be?	6. Rico has 4 dimes. What could the other coins be?
7. Olivia has no dimes or pennies. What could the coins be?	8. Dillon has no nickels or pennies. What could the coins be?	9. Theodore has no dimes. What could the coins be?

Name ______________________________ 2.MD.C.8

Money Problem Solving

Use information from the menu to make a math problem. Solve the problem. Read each story carefully and decide if it makes sense to add or subtract.

Smokey Joe's Barbecue

MAIN DISHES		SIDE DISHES		BEVERAGES	
Eye-Watering Ham	\$3.50	Flame Fries	\$1.10	Cola	\$0.75
Burning-Hot Ribs	\$3.75	Sizzlin' Salad	\$1.05	Lemonade	\$0.85
Rockin' Roast Beef	\$4.25	Tasty Tater Tots	\$0.95	Milk	\$0.95

1. Ariel ordered ribs and lemonade. How much will her lunch cost?

2. Michael ordered roast beef. He paid with a five-dollar bill. How much change will he get?

3. Jonah has \$4.08. He buys ham as a main dish. How much money does Jonah have left?

4. Terone wonders, "How much does an order of ribs, fries, and a cola cost?"

5. How much more does roast beef cost than milk?

6. Kelsey orders the least expensive item from each section of the menu. How much does she spend?

7. Ryan spent \$5.55 for lunch. He got \$0.45 back as change. How much money did Ryan start out with?

8. Tracy buys lemonade for herself and three friends. How much does she spend?

Name ______________________________ 2.MD.D.9

Line Plots

A line plot uses a number line and Xs to show data that has been collected.

The line plot shows the height in feet of the sunflowers in Ms. Park's garden. Read the graph and answer the questions.

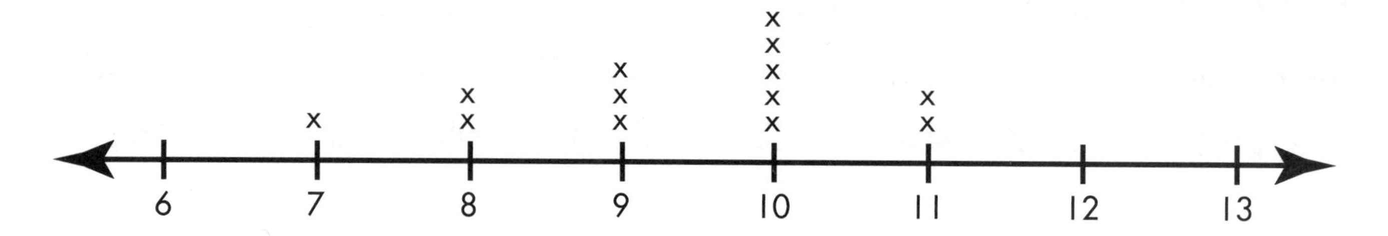

1. What is the most common height of the sunflowers?

2. How many sunflowers are 10 feet tall?

3. How many sunflowers are 8 feet tall?

4. Which height shows three sunflowers?

5. Ms. Park measured two more sunflowers. The first one was 8 feet tall and the second one was 12 feet tall. Mark **X**s on the line plot to add the sunflowers to the graph.

Name ______________________ 2.MD.D.9

Line Plots

Brooke made necklaces of different lengths to sell at the school carnival.

Lengths of Necklaces

16 inches	18 inches
20 inches	17 inches
16 inches	16 inches
17 inches	19 inches
18 inches	16 inches

Use the data to complete the line plot. Answer the questions.

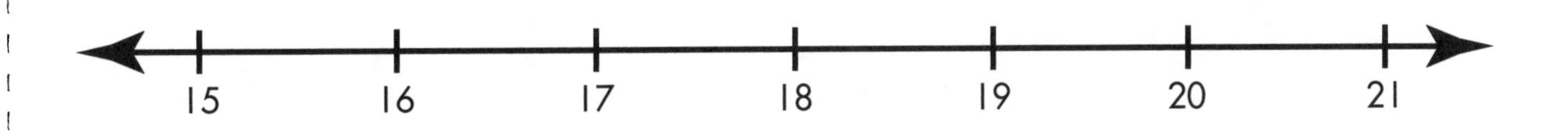

1. What was the total number of 16-inch necklaces?

2. What was the total number of 18-inch necklaces?

3. Brooke made one more necklace that was 20 inches long. Graph that necklace on the line plot.

4. What was the total number of necklaces that Brooke made?

Name ______________________ (2.MD.D.9)

Line Plots

Mrs. Rivera's class planted beans. After a week, the class recorded the height of the sprouts. Use the following data to make a line plot in the space provided. Answer the questions.

Bean Sprouts

Height in centimeters	Number they counted
3 cm	4
4 cm	7
5 cm	6
6 cm	7
7 cm	3

1. Which two heights had the same number of bean sprouts?

2. How many bean sprouts were 7 centimeters in height?

3. What was the total number of bean sprouts?

4. How many more bean sprouts were 5 centimeters than 3 centimeters in height?

Name ______________________________ 2.MD.D.10

Interpreting Graphs

Graphs can be used to observe and compare information. **Picture graphs**, or **pictographs**, often have a key that will tell what each picture means. In this graph, one circle equals two students.

Margie's class made a frequency table to show what the second graders did during choice time. Use the table to make a pictograph below. The finished graph will help you answer the questions.

Activity	Number of Students
read	18
finish work	6
dice math	24
color	12
clean desk	6
science project	18

Draw 1 ○ for every 2 students.

Activity	Number of Students
read	
finish work	
dice math	
color	
clean desk	
science project	

○ = 2 students

1. If 10 students made an art project, how many circles would you draw? ______________

2. If each ○ = 2 students,
   how would you show 1 student? _____ 9 students? ______________

3. Circle **T** for true or **F** for false.

   T F More students chose dice math than coloring.
   T F An equal number of students read as did the science project.
   T F Fewer students colored than cleaned desks.
   T F Ten more students finished work than read.

Name ______________________________ 2.MD.D.10

Interpreting Graphs

Use the data to complete the bar graph. Starting at the bottom, color a section for every candy that Mrs. Nickles passed out. The finished graph will help you answer the questions.

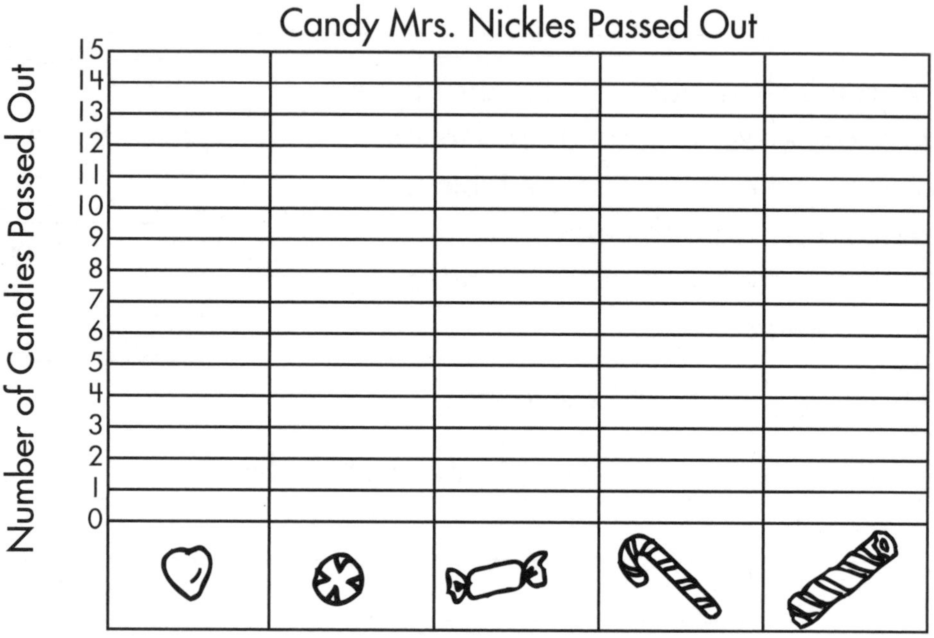

Candy Mrs. Nickles Passed Out

Number of Candies Passed Out

15
14
13
12
11
10
9
8
7
6
5
4
3
2
1
0

1. Which candy totaled 11 pieces? ______

2. Which two candies combined total 23 pieces? ______ and ______

3. How many more pieces of  than  are there? ______

4. How many pieces of  and  together did Mrs. Nickles give away? ______

5. Which two candies combined total 21 pieces? ______ and ______

6. How many more pieces of  than  are there? ______

7. What is the total number of pieces of  ,  , and ? ______

Name ______________________ 2.MD.D.10

Interpreting Graphs

Use the information from the graphs to answer each question.

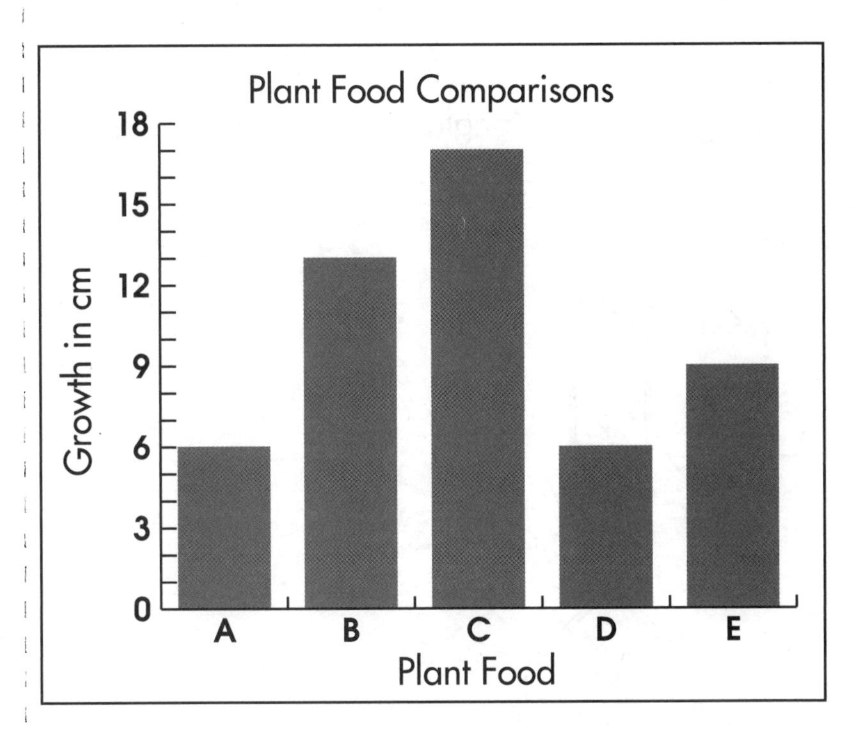

1. How much did Food B's plant grow? ______________

2. How much more did Food C's plant grow than Food A's plant? ______

3. Which two plants show the same growth? ______ and ______

Flowers We Planted

Rm 102	○ ○ ○ ● ● ●
Rm 103	○ ○ ● ● ●
Rm 104	○ ● ● ● ● ●
Rm 105	○ ○ ● ● ● ●

○ = sun-loving flowers ● = shade-loving flowers

4. Which two rooms planted the same number of shade-loving flowers? ______ and ______

5. How many sun-loving flowers were planted? ______________

6. Which room planted the same number of each flower? ______

Name ______________________ 2.G.A.1

Identifying Shapes

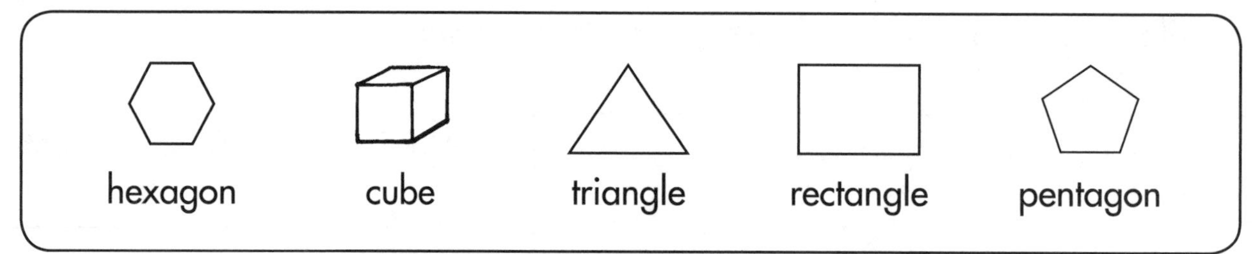

Write the name of the shape next to each item.

1.  ______________

2.  ______________

3.  ______________

4.  ______________

5.  ______________

6.  ______________

7.  ______________

8.  ______________

Name ______________________________ 2.MD.D.10, 2.G.A.1

Identifying Shapes

Shapes that are 3-dimensional are called **solid**. A round, solid shape is not called a circle; it is called a sphere. Look at the solid figures and their names in the chart.

Count each solid figure in the picture below. Then, color the boxes in the graph to show how many of each figure you found.

6					
5					
4					
3					
2					
1					
	Sphere	Cube	Cone	Rectangular Prism	Cylinder

Name ______________________________ 2.G.A.1

Identifying Shapes

A **solid figure** has length, width, height, and volume.

A **face** is a flat side of a solid figure.

Connect each figure with its name. Then, connect each name with the number of faces it has.

	Figure	Name	Number of Faces
1.		pyramid	1
2.		cube	6
3.		cylinder	5
4.		cone	0
5.		sphere	2
6.		rectangular prism	6

Classify each item by its shape. Write its name in the chart.

marble

party hat

cereal box

globe

straw

block

die

candle

pyramid

soccer cone

Cube	Rectangular Prism
Pyramid	**Cone**
Cylinder	**Sphere**

Name ______________________ 2.G.A.1

Shape Attributes

A **polygon** is a closed plane figure made of line segments.

A polygon with three sides is a **triangle**.

A polygon with four sides is a **quadrilateral**.

A polygon with five sides is a **pentagon**.

Use the code to color each shape.

Color Code	
triangles:	red
quadrilaterals:	green
pentagons:	blue

1.  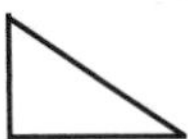  

2. 

3. 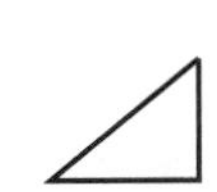 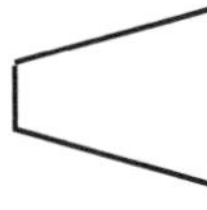 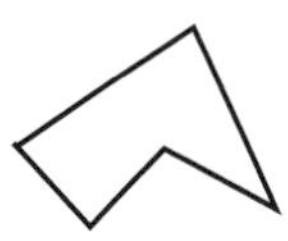 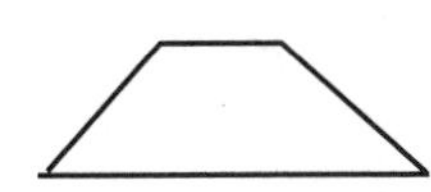

A **parallelogram** is a quadrilateral with two pairs of opposite parallel sides.

Finish each parallelogram pattern.

4.								
5.								
6.			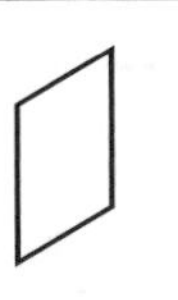	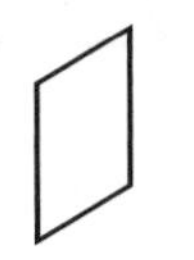	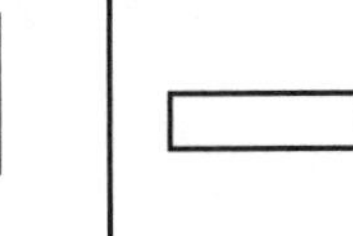 	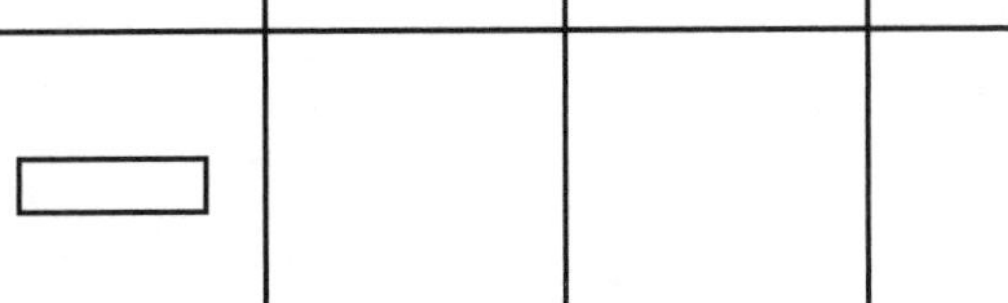		

Name ______________________ 2.G.A.1

Shape Attributes

A **parallelogram** is a quadrilateral with opposite sides parallel. Opposite sides and angles are congruent.

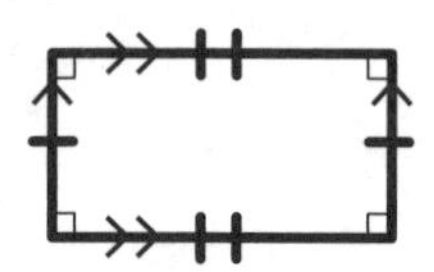

A **rectangle** is a parallelogram with four right angles. Opposite sides are congruent and parallel.

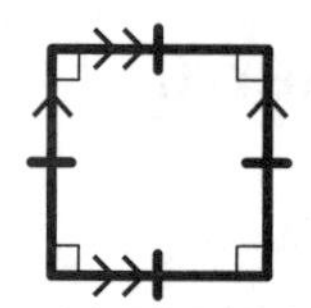

A **square** is a rectangle with four congruent sides. Opposite sides are parallel.

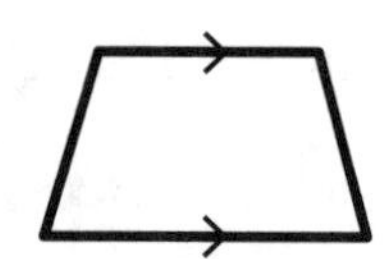

A **trapezoid** is a quadrilateral with exactly one pair of parallel sides.

A **rhombus** is a parallelogram with four congruent sides. Opposite angles are congruent and opposite sides are parallel.

Name each figure.

1. 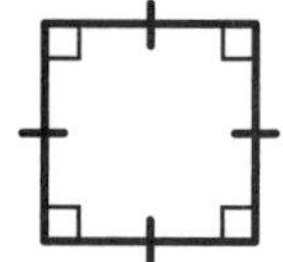

____

2. 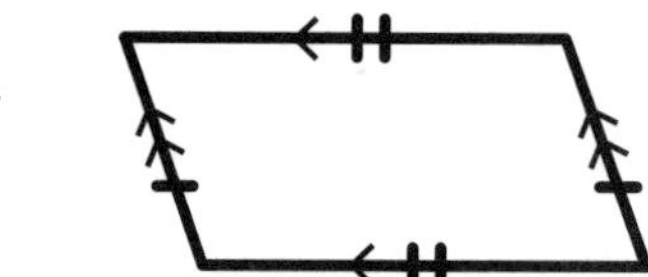

____

3. 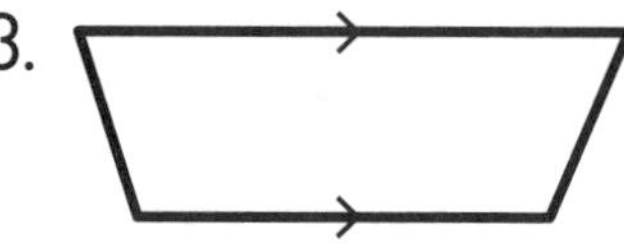

____

4. 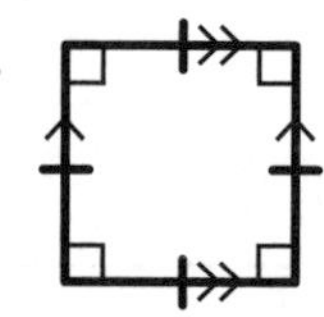

____

5. 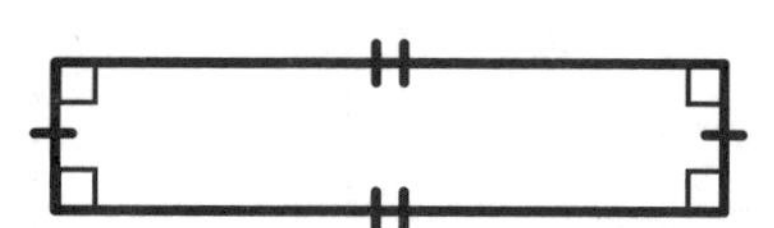

____

6. 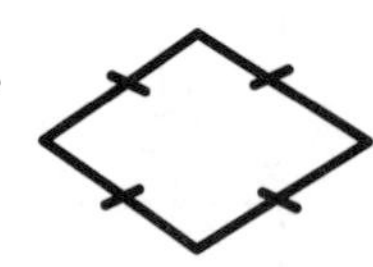

____

7. 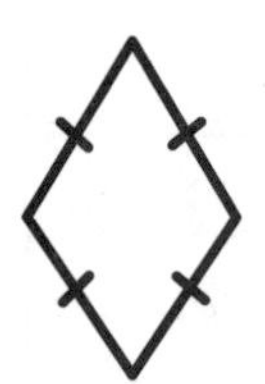

____

8. 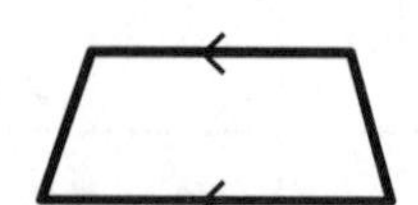

____

Name ______________________________ (2.G.A.1)

Shape Attributes

A **quadrilateral** is a closed figure with four sides and four angles. Make four different quadrilaterals on your geoboard. Record your figures here.

1. 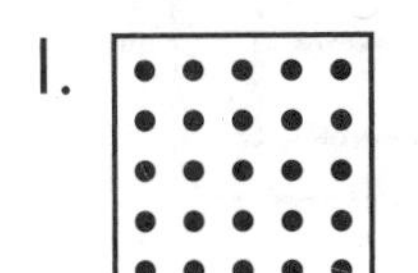
2. 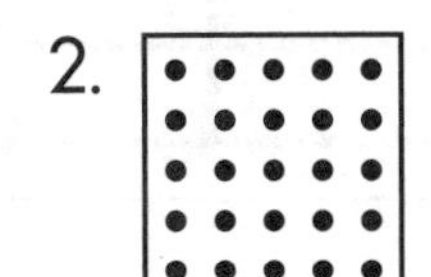
3. 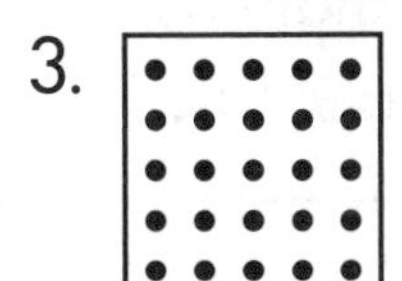
4. 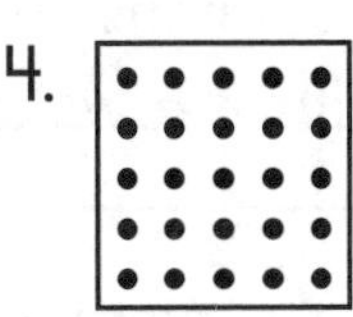

What makes each of these figures a quadrilateral? ______________________________

There are special types of quadrilaterals.

- A **trapezoid** is a quadrilateral with just one set of parallel sides.
- A **parallelogram** is a quadrilateral with two sets of parallel sides.
- A **rectangle** is a parallelogram with four right angles.
- A **square** is a rectangle with four sides of equal length.

Use a geoboard to make each figure described. Record them on the grids below.

5. 
a trapezoid

6. 
a parallelogram

7. 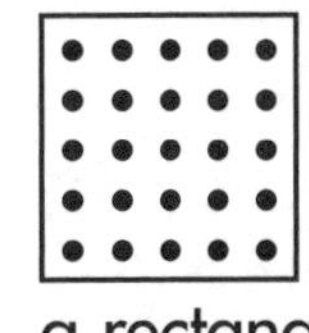
a rectangle

8. 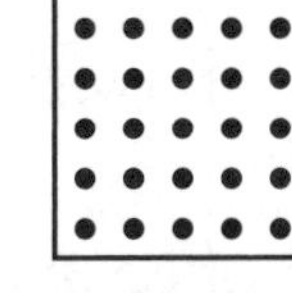
a square

9. 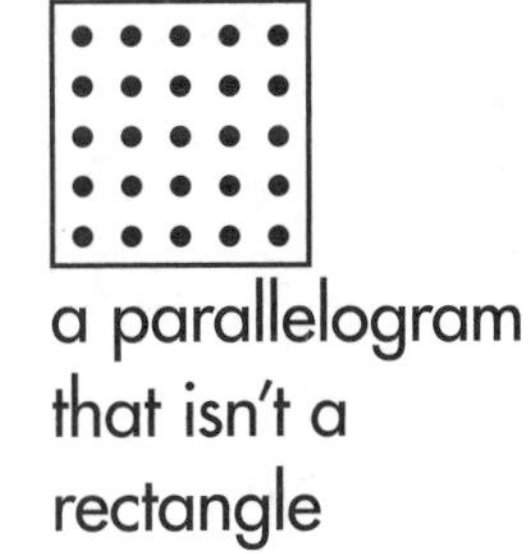
a parallelogram that isn't a rectangle

10. 
a parallelogram that is a square

11. 
a quadrilateral that is a trapezoid

12. 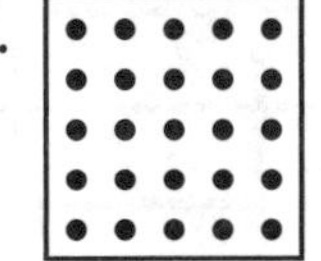
a rectangle that is a square

13. 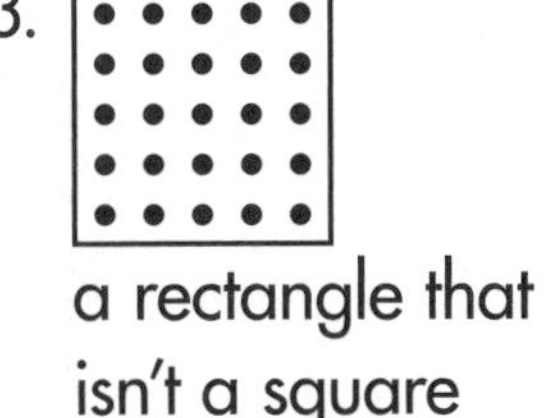
a rectangle that isn't a square

14. 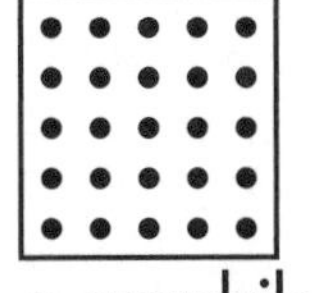
a quadrilateral that isn't a trapezoid or parallelogram

15. 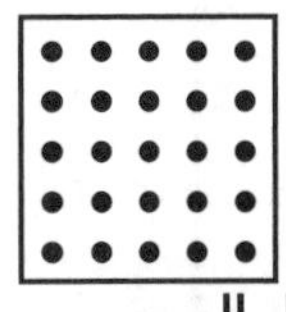
a parallelogram that is a rectangle

16. 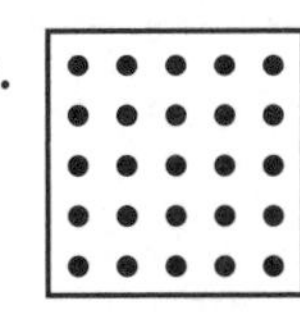
a square that is a rectangle

Name ______________________________ 2.G.A.2

Partitioning Rectangles

You can **partition** rectangles into equal parts using rows and columns. This rectangle has 9 equal parts with 3 rows and 3 columns.

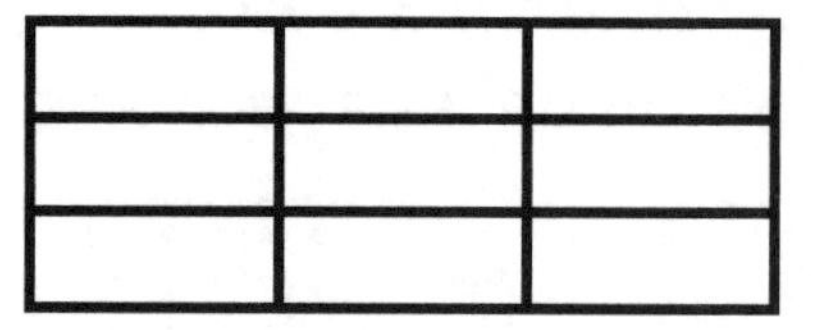

Look at each rectangle. Answer the questions.

1. Number of rows? _____

   Number of columns? _____

2. Number of rows? _____

   Number of columns? _____

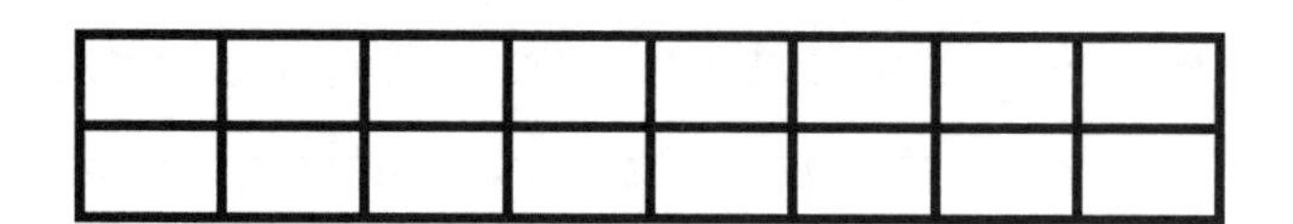

3. Number of rows? _____

   Number of columns? _____

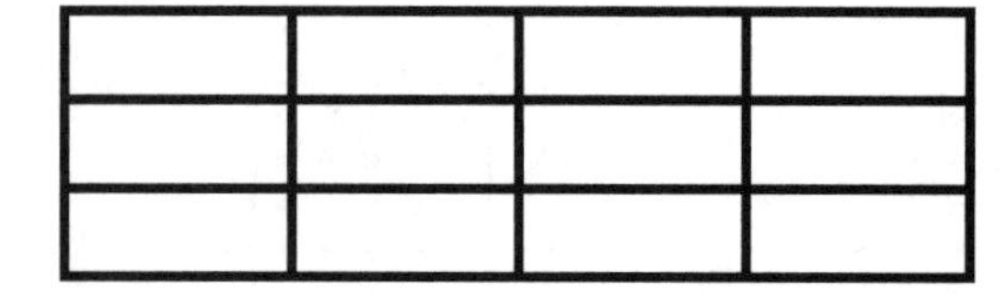

4. Partition this rectangle equally into 6 columns and 2 rows.

5. Partition this rectangle equally into 4 rows and 5 columns.

6. Partition this rectangle equally into 5 columns and 5 rows.

Name ______________________________ 2.G.A.2

Partitioning Rectangles

Use the grid to make a rectangle with:

1. 6 rows and 3 columns

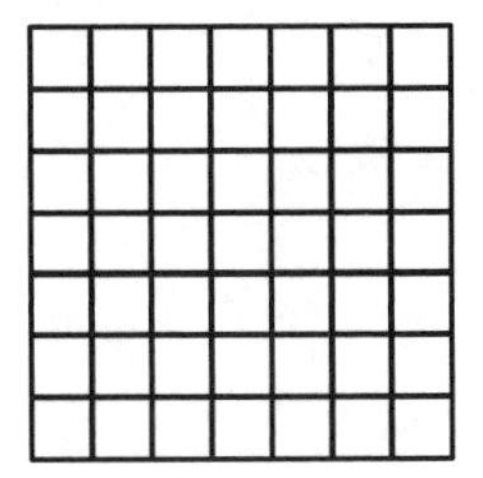

2. 1 row and 6 columns

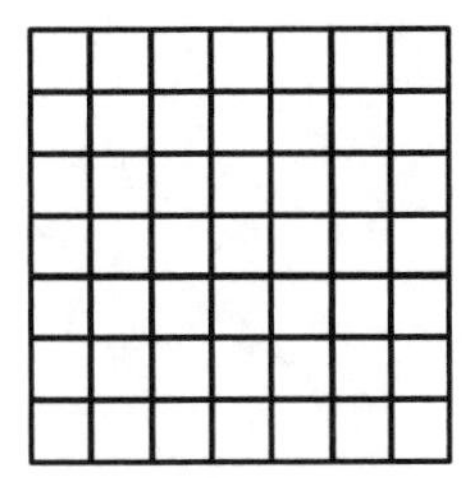

3. 4 rows and 4 columns

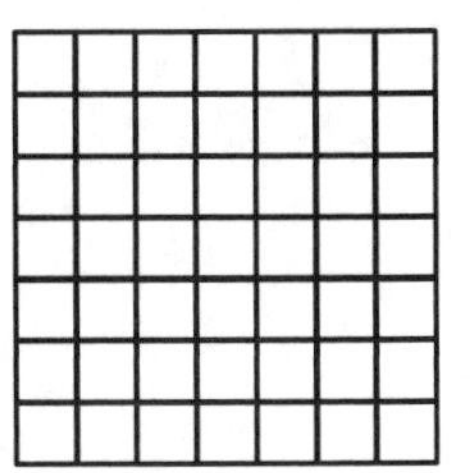

4. 5 rows and 3 columns

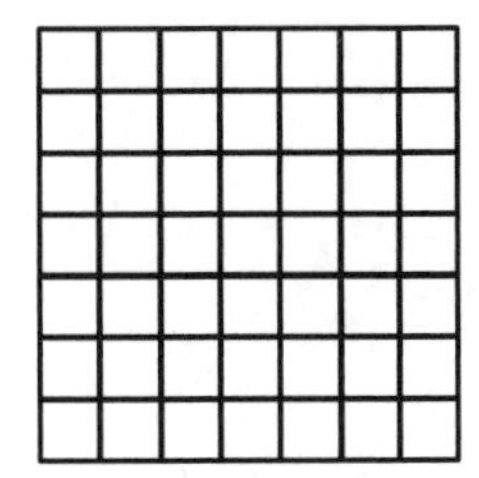

5. Partition this rectangle equally into 7 columns and 2 rows.

How many parts did you draw? _______

6. Partition this rectangle equally into 4 rows and 6 columns.

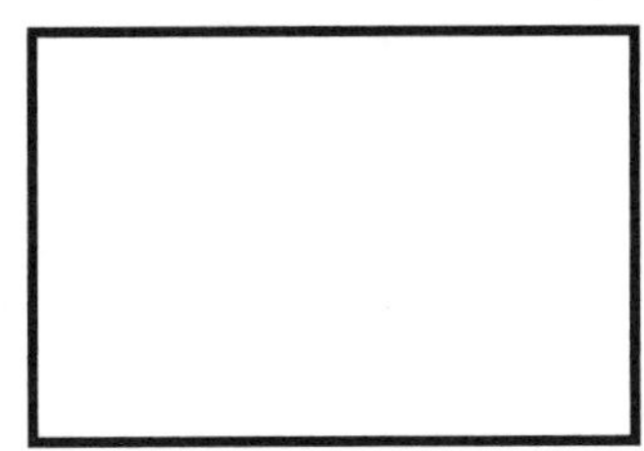

How many parts did you draw? _______

Name ______________________________ (2.G.A.2)

Partitioning Rectangles

Using color tiles or square pattern blocks, draw and record as many different rectangles as you can with the given number.

1. 8 squares

2. 10 squares

3. 12 squares

Name ______________________________ 2.G.A.3

Equal Parts of Shapes

Equal parts mean pieces that are exactly the same.

equal parts: not equal parts: 

When something is divided into equal parts, we can call the parts by special names. half third fourth

Two equal parts are called halves. Three equal parts are called thirds. Four equal parts are called fourths.

Color the stars that show equal parts.

Name ______________________________ 2.G.A.3

Equal Parts of Shapes

Shade to show the equal shares.

1. one-fourth

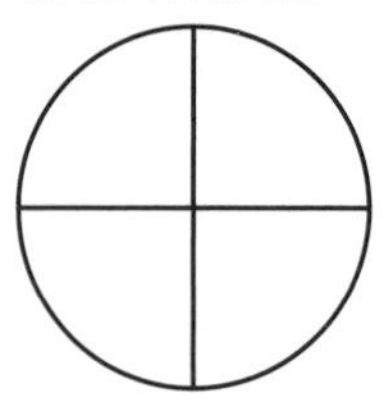

2. two-fourths

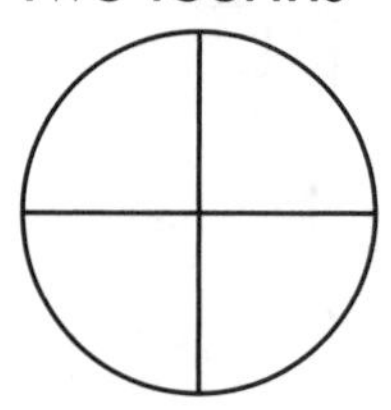

3. two-thirds

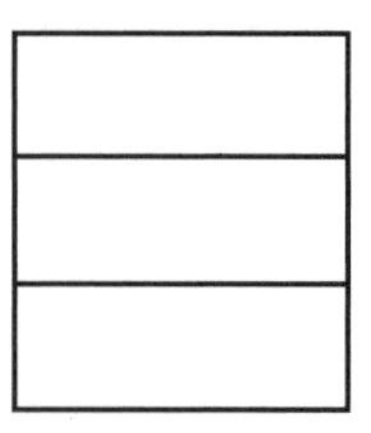

4. two-fourths

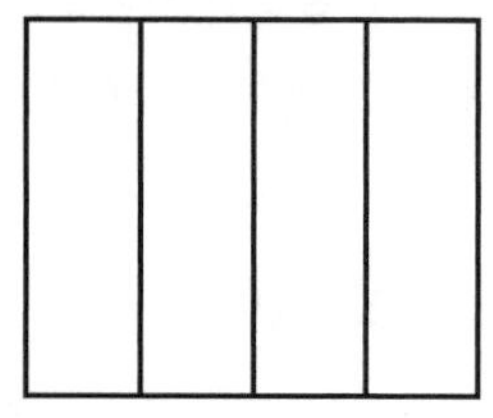

5. three-fourths

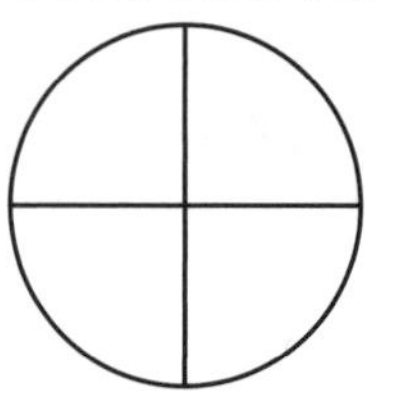

6. one-half

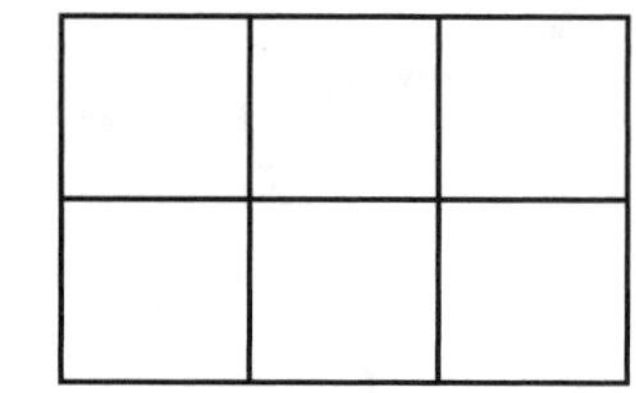

7. one-third

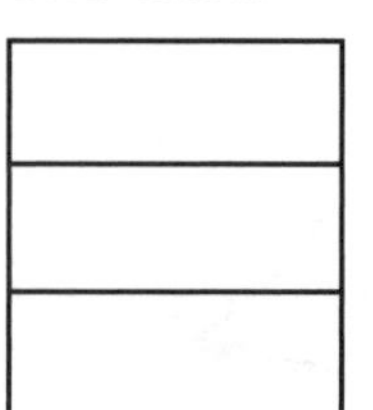

8. one-third

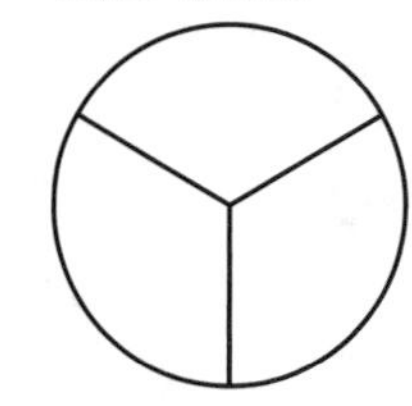

9. Show 3 equal shares.

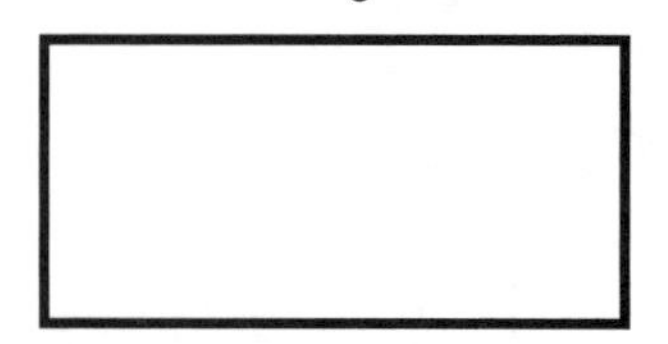

10. Show 4 equal shares.

11. Dylan's mother wanted to cut his birthday cake into 4 equal shares. Show three ways she could do this.

Name ______________________ 2.G.A.3

Equal Parts of Shapes

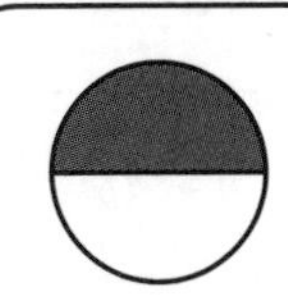

$\frac{1}{2}$ The top number in a fraction tells how many parts are shaded.
The bottom number in a fraction tells how many equal parts.

Write the fraction for the shaded part.

1. 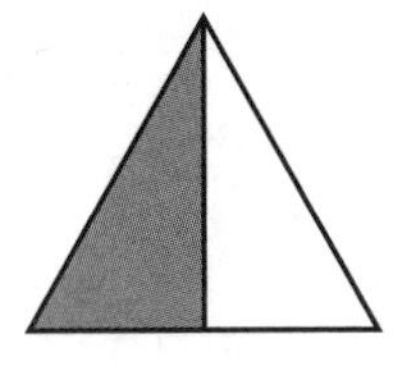 $\frac{\square}{\square}$

2. 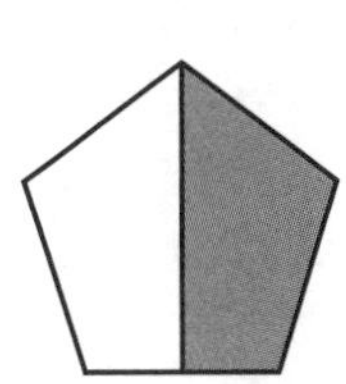 $\frac{\square}{\square}$

3. 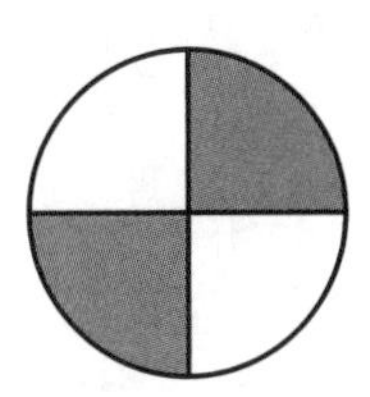 $\frac{\square}{\square}$

4. 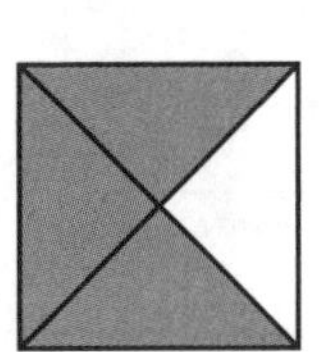 $\frac{\square}{\square}$

5. 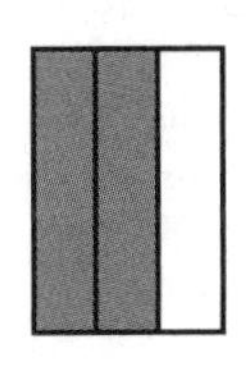 $\frac{\square}{\square}$

6. 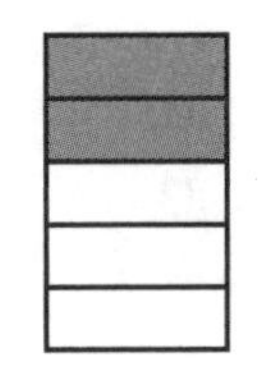 $\frac{\square}{\square}$

7. 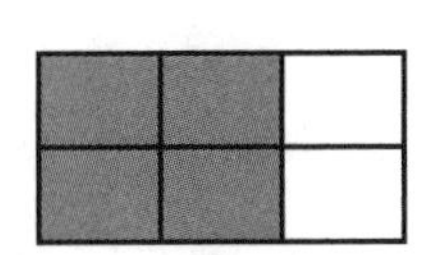 $\frac{\square}{\square}$

8. 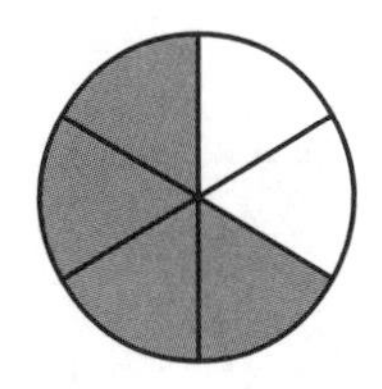 $\frac{\square}{\square}$

Answer Key

Name ____________ (2.OA.A.1)

Fact Families

Some addition and subtraction problems are related, like families. They can make two addition and two subtraction problems using the same three numbers.

5 + 6 = 11
6 + 5 = 11
11 − 5 = 6
11 − 6 = 5

Complete each fact family.

1. 2 + 8 = 10 **8** + 2 = 10 10 − 8 = 2 10 − 2 = **8**	2. 5 + 4 = 9 4 + **5** = 9 9 − 5 = **4** 9 − 4 = **5**	3. 6 + 9 = 15 9 + 6 = **15** 15 − 9 = 6 **15** − 6 = 9
4. 5 + 7 = 12 7 + **5** = 12 **12** − 5 = 7 **12** − 7 = 5	5. 3 + 9 = 12 **9** + 3 = 12 12 − 9 = 3 **12** − 3 = **9**	6. 7 + 9 = 16 9 + **7** = 16 16 − **9** = **7** **16** − 7 = 9

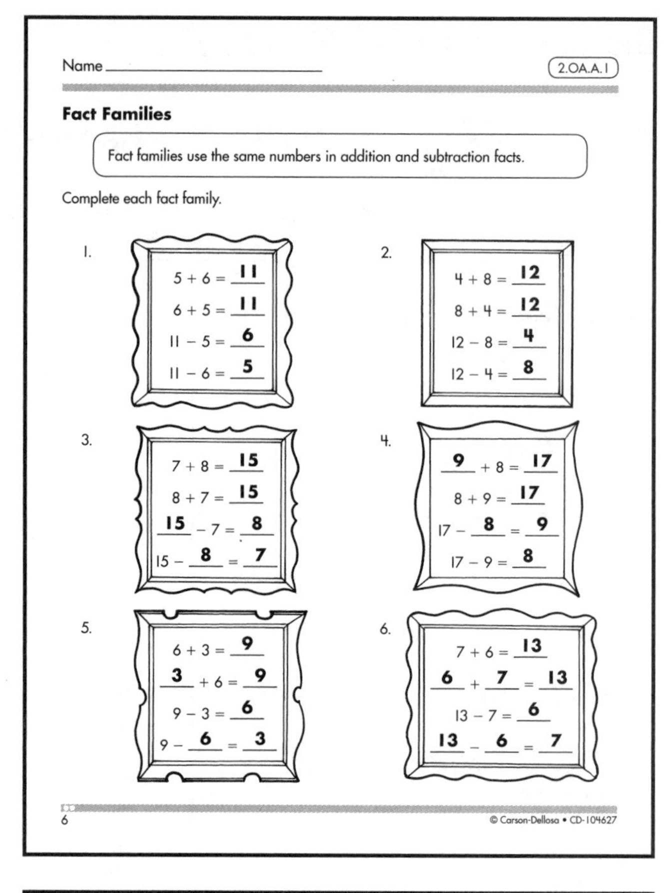
Name ____________ (2.OA.A.1)

Fact Families

Fact families use the same numbers in addition and subtraction facts.

Complete each fact family.

1.
5 + 6 = **11**
6 + 5 = **11**
11 − 5 = **6**
11 − 6 = **5**

2.
4 + 8 = **12**
8 + 4 = **12**
12 − 8 = **4**
12 − 4 = **8**

3.
7 + 8 = **15**
8 + 7 = **15**
15 − 7 = **8**
15 − **8** = **7**

4.
9 + 8 = **17**
8 + 9 = **17**
17 − **8** = **9**
17 − 9 = **8**

5.
6 + 3 = **9**
3 + 6 = **9**
9 − 3 = **6**
9 − **6** = **3**

6.
7 + 6 = **13**
6 + **7** = **13**
13 − 7 = **6**
13 − **6** = **7**

Name ____________ (2.OA.A.1)

Fact Families

Complete each fact family.

1. 8 14 6 **8** + **6** = **14** **6** + **8** = **14** **14** − **8** = **6** **14** − **6** = **8**	2. 5 12 7 **5** + **7** = **12** **7** + **5** = **12** **12** − **7** = **5** **12** − **5** = **7**
3. 8 11 3 **8** + **3** = **11** **3** + **8** = **11** **11** − **8** = **3** **11** − **3** = **8**	4. 7 15 8 **7** + **8** = **15** **8** + **7** = **15** **15** − **8** = **7** **15** − **7** = **8**
5. 7 13 6 **7** + **6** = **13** **6** + **7** = **13** **13** − **6** = **7** **13** − **7** = **6**	6. 9 17 8 **9** + **8** = **17** **8** + **9** = **17** **17** − **9** = **8** **17** − **8** = **9**

Name ____________ (2.OA.A.1)

Number Sentences

Each part of an addition problem has a name. The numbers being added are called **addends** and the answer is called the **sum**. They form a number sentence. The addends can be switched around, and the sum stays the same! This is called the **commutative property of addition**.

Example: 8 + 7 = 15 (8 and 7: addends; 15: sum) 7 + 8 = 15 So . . . 8 + 7 = 7 + 8

Fill in the missing addends and sums.

1. 6 + **4** = 10
2. 5 + **8** = 13
3. 7 + **5** = 12
4. **7** + 7 = 14
5. **2** + 9 = 11
6. **2** + 8 = 10
7. 8 + **4** = 12
8. 4 + **6** = 10
9. 9 + **6** = 15
10. **5** + 5 = 10
11. **1** + 8 = 9
12. **6** + 7 = 13
13. 7 + **4** = 11
14. 6 + **6** = 12
15. 9 + **5** = 14
16. 8 + 7 = **7** + 8
17. **9** + 5 = 5 + 9

Answer Key

Name ______________________ (2.OA.A.1)

Number Sentences

Fill in the missing addends and sums.

1. 6 + **10** = 16
2. 7 + **6** = 13
3. 10 + **2** = 12
4. **4** + 7 = 11
5. **9** + 9 = 18
6. **9** + 8 = 17
7. 8 + **10** = 18
8. 4 + **9** = 13
9. 9 + **5** = 14
10. **10** + 5 = 15
11. **6** + 8 = 14
12. **9** + 7 = 16
13. 7 + **9** = 16
14. 9 + **3** = 12
15. 10 + **4** = 14
16. **8** + 9 = 17
17. **3** + 8 = 11
18. **8** + 7 = 15
19. 8 + 7 = **7** + 8
20. **9** + 7 = 7 + 9
21. 6 + **9** = 9 + 6
22. 7 + 6 = 6 + **7**

Name ______________________ (2.OA.A.1)

Number Sentences

Fill in the missing addends and sums.

1. 6 + **12** = 18
2. 5 + **11** = 16
3. 7 + **12** = 19
4. **10** + 7 = 17
5. **11** + 9 = 20
6. **12** + 8 = 20
7. 8 + **7** = 15
8. 4 + **9** = 13
9. 9 + **10** = 19
10. **15** + 5 = 20
11. **8** + 8 = 16
12. **13** + 7 = 20
13. 7 + **12** = 19
14. 6 + **14** = 20
15. 9 + **9** = 18
16. **9** + 9 = 18
17. **3** + 8 = 11
18. **7** + 7 = 14
19. 18 + 17 = **17** + 18
20. **9** + 15 = 15 + 9
21. 60 + **90** = 90 + 60
22. 70 + 60 = 60 + **70**

Name ______________________ (2.OA.A.1)

Using Number Sentences in Word Problems

Use these steps to solve word problems:

1. Read the story and the question.
2. Read the question again.
3. Circle the numbers in the story that you need to answer the question.
4. Watch for key words: *altogether, how many more, in all.*
5. Choose to **+** or **−**.
6. Answer the question.
7. Use pictures, words, or numbers to show your work.

Solve each problem with a number sentence. Show your work with pictures, words, or numbers.

Check students' work to verify answers.

1. Megan has 10 baseball cards. Taylor has 14 baseball cards. How many more baseball cards does Taylor have than Megan?

   10 + **4** = 14 or 14 − 10 = **4** baseball cards

2. Matt has 9 crayons. Josh has 8 more crayons than Matt. How many crayons does Josh have?
   17 crayons
3. Kevin has 11 more toy cars than Luke. Kevin has 16 toy cars. How many toy cars does Luke have?
   5 toy cars
4. A team has 12 students. Nine of the students are girls. How many students are boys?
   3 boys
5. A bus has 15 students riding on it. At the first bus stop, seven students get off. How many students are left on the bus?
   8 students

Name ______________________ (2.OA.A.1)

Using Number Sentences in Word Problems

Solve the problem with a number sentence. Show your work with pictures, words, or numbers.

Check students' work to verify answers.

1. Megan has 20 baseball cards. Taylor has 42 baseball cards. How many more baseball cards does Taylor have?

   20 + **22** = 42 or 42 − 20 = **22**

2. Matt has 29 crayons. Josh has 11 more than Matt. How many crayons does Josh have?
   40 crayons
3. Kevin has 31 more toy cars than Luke. Kevin has 51 toy cars. How many toy cars does Luke have?
   20 toy cars
4. There are 26 students on a team. Fifteen of the students are girls. How many students are boys?
   11 boys
5. There are 39 students on a bus. Eighteen students get off the bus at the first stop. How many students are left on the bus?
   21 students
6. There are 18 crackers on a plate. Miquel ate 4 crackers. Then Nathan ate 2 crackers. How many crackers are left on the plate?
   12 crackers

Answer Key

Name ____________________ (2.OA.A.1)

Using Number Sentences in Word Problems

Solve the problem with a number sentence. Show your work with pictures, words, or numbers.

Check students' work to verify answers.

1. Megan has 76 baseball cards. Taylor has 89 baseball cards. How many more baseball cards does Taylor have?
   13 baseball cards
2. Matt has 65 crayons. Josh has 33 more than Matt. How many crayons does Josh have?
   98 crayons
3. Kevin has 21 more toy cars than Luke. Kevin has 61 toy cars. How many toy cars does Luke have?
   40 toy cars
4. There are 45 students on a team. Twenty-three of the students are girls. How many students are boys?
   22 boys
5. There are 69 students on a bus. Seventeen students get off the bus at the first stop. How many students are left on the bus?
   52 students
6. There are 32 crackers on a plate. Miquel ate 16 crackers. Then Nathan ate 7 crackers. How many crackers are left on the plate?
   9 crackers

Name ____________________ (2.OA.B.2)

Addition Fluency

Solve each problem.

1. 7 + 6 **13**	2. 7 + 7 **14**	3. 3 + 6 **9**	4. 4 + 3 **7**	5. 7 + 4 **11**	6. 5 + 4 **9**
7. 6 + 5 **11**	8. 9 + 2 **11**	9. 8 + 4 **12**	10. 6 + 2 **8**	11. 3 + 2 **5**	12. 3 + 9 **12**
13. 3 + 3 **6**	14. 9 + 5 **14**	15. 9 + 6 **15**	16. 2 + 7 **9**	17. 6 + 4 **10**	18. 5 + 8 **13**
19. 3 + 7 **10**	20. 6 + 8 **14**	21. 8 + 3 **11**	22. 5 + 4 **9**	23. 9 + 1 **10**	24. 2 + 8 **10**
25. 2 + 3 **5**	26. 3 + 5 **8**	27. 4 + 9 **13**	28. 4 + 3 **7**	29. 8 + 8 **16**	30. 3 + 3 **6**

Name ____________________ (2.OA.B.2)

Addition Fluency

Solve each problem.

1. 6 + 3 **9**	2. 7 + 3 **10**	3. 5 + 5 **10**	4. 8 + 2 **10**	5. 8 + 3 **11**	6. 6 + 4 **10**
7. 5 + 3 **8**	8. 3 + 4 **7**	9. 6 + 6 **12**	10. 3 + 3 **6**	11. 9 + 0 **9**	12. 7 + 3 **10**
13. 9 + 3 **12**	14. 9 + 2 **11**	15. 5 + 4 **9**	16. 1 + 2 **3**	17. 0 + 9 **9**	18. 8 + 4 **12**
19. 4 + 4 **8**	20. 7 + 5 **12**	21. 3 + 8 **11**	22. 8 + 4 **12**	23. 3 + 7 **10**	24. 7 + 2 **9**
25. 6 + 2 **8**	26. 5 + 6 **11**	27. 5 + 1 **6**	28. 4 + 5 **9**	29. 9 + 2 **11**	30. 6 + 6 **12**

Name ____________________ (2.OA.B.2)

Addition Fluency

Solve each problem.

1. 9 + 5 **14**	2. 7 + 7 **14**	3. 8 + 5 **13**	4. 6 + 7 **13**	5. 5 + 5 **10**	6. 6 + 4 **10**
7. 9 + 7 **16**	8. 9 + 4 **13**	9. 5 + 6 **11**	10. 9 + 9 **18**	11. 6 + 3 **9**	12. 7 + 4 **11**
13. 8 + 6 **14**	14. 7 + 5 **12**	15. 9 + 7 **16**	16. 9 + 5 **14**	17. 8 + 7 **15**	18. 9 + 3 **12**
19. 8 + 2 **10**	20. 9 + 4 **13**	21. 8 + 8 **16**	22. 9 + 3 **12**	23. 8 + 3 **11**	24. 9 + 8 **17**
25. 6 + 6 **12**	26. 7 + 3 **10**	27. 9 + 9 **18**	28. 14 + 2 **16**	29. 16 + 2 **18**	30. 17 + 1 **18**

Answer Key

Name ______________________ (2.OA.B.2)

Subtraction Fluency

Solve each problem.

1. 9	2. 8	3. 10	4. 10	5. 10	6. 9
− 6	− 2	− 10	− 6	− 7	− 6
3	**6**	**0**	**4**	**3**	**3**

7. 7	8. 7	9. 10	10. 10	11. 9	12. 10
− 2	− 4	− 4	− 0	− 4	− 2
5	**3**	**6**	**10**	**5**	**8**

13. 8	14. 10	15. 7	16. 9	17. 8	18. 8
− 6	− 2	− 5	− 5	− 5	− 7
2	**8**	**2**	**4**	**3**	**1**

19. 7	20. 8	21. 10	22. 6	23. 10	24. 9
− 3	− 4	− 3	− 2	− 1	− 3
4	**4**	**7**	**4**	**9**	**6**

25. 7	26. 10	27. 10	28. 10	29. 8	30. 10
− 6	− 4	− 8	− 5	− 1	− 9
1	**6**	**2**	**5**	**7**	**1**

© Carson-Dellosa • CD-104627 17

Name ______________________ (2.OA.B.2)

Subtraction Fluency

Solve each problem.

1. 10	2. 15	3. 13	4. 15	5. 12	6. 14
− 9	− 9	− 5	− 8	− 7	− 6
1	**6**	**8**	**7**	**5**	**8**

7. 10	8. 12	9. 13	10. 15	11. 12	12. 15
− 4	− 8	− 9	− 5	− 5	− 3
6	**4**	**4**	**10**	**7**	**12**

13. 14	14. 12	15. 12	16. 11	17. 15	18. 10
− 5	− 6	− 4	− 6	− 2	− 6
9	**6**	**8**	**5**	**13**	**4**

19. 13	20. 13	21. 12	22. 13	23. 15	24. 14
− 6	− 8	− 9	− 3	− 7	− 7
7	**5**	**3**	**10**	**8**	**7**

25. 15	26. 10	27. 15	28. 12	29. 15	30. 12
− 8	− 5	− 1	− 1	− 2	− 6
7	**5**	**14**	**11**	**13**	**6**

18 © Carson-Dellosa • CD-104627

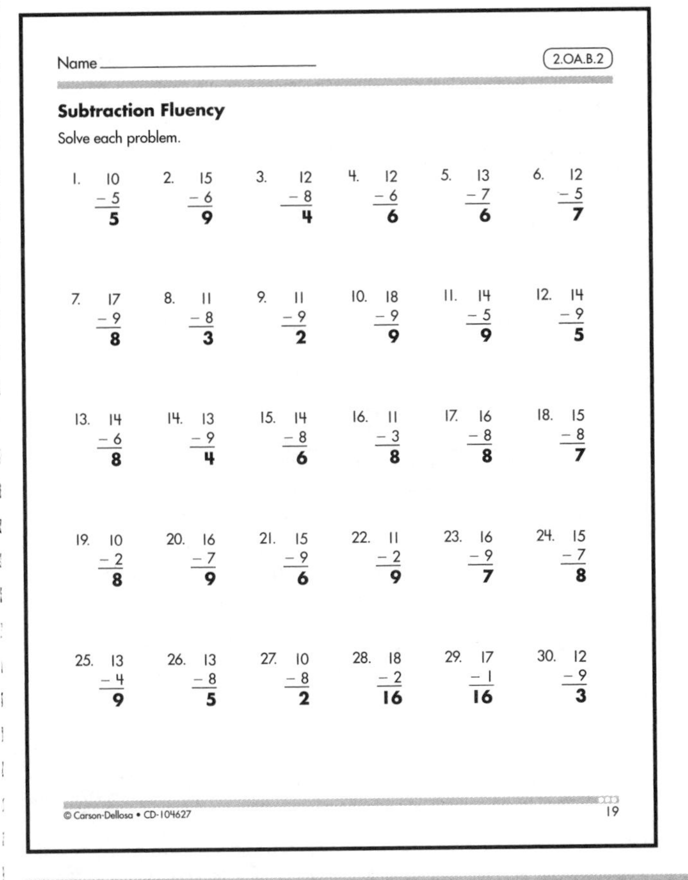

Name ______________________ (2.OA.B.2)

Subtraction Fluency

Solve each problem.

1. 10	2. 15	3. 12	4. 12	5. 13	6. 12
− 5	− 6	− 8	− 6	− 7	− 5
5	**9**	**4**	**6**	**6**	**7**

7. 17	8. 11	9. 11	10. 18	11. 14	12. 14
− 9	− 8	− 9	− 9	− 5	− 9
8	**3**	**2**	**9**	**9**	**5**

13. 14	14. 13	15. 14	16. 11	17. 16	18. 15
− 6	− 9	− 8	− 3	− 8	− 8
8	**4**	**6**	**8**	**8**	**7**

19. 10	20. 16	21. 15	22. 11	23. 16	24. 15
− 2	− 7	− 9	− 2	− 9	− 7
8	**9**	**6**	**9**	**7**	**8**

25. 13	26. 13	27. 10	28. 18	29. 17	30. 12
− 4	− 8	− 8	− 2	− 1	− 9
9	**5**	**2**	**16**	**16**	**3**

© Carson-Dellosa • CD-104627 19

Name ______________________ (2.OA.B.2)

Addition and Subtraction Fluency

Solve each problem.

1. 6 + 3 = **9**	2. 8 − 6 = **2**	3. 5 + 3 = **8**
4. 10 − 10 = **0**	5. 9 + 0 = **9**	6. 9 − 0 = **9**
7. 8 + 1 = **9**	8. 8 + 3 = **11**	9. 7 − 2 = **5**
10. 5 + 1 = **6**	11. 9 − 2 = **7**	12. 5 + 4 = **9**
13. 6 − 4 = **2**	14. 7 + 3 = **10**	15. 8 − 2 = **6**
16. 2 + 6 = **8**	17. 10 − 6 = **4**	18. 10 − 1 = **9**
19. 6 + 4 = **10**	20. 8 − 2 = **6**	21. 2 + 7 = **9**
22. 8 − 5 = **3**	23. 5 + 5 = **10**	24. 5 − 5 = **0**

20 © Carson-Dellosa • CD-104627

Answer Key

Name ______________________ (2.OA.B.2)

Addition and Subtraction Fluency

Solve each problem.

1. 18 − 9 = **9**
2. 10 + 5 = **15**
3. 13 − 7 = **6**
4. 14 + 6 = **20**
5. 15 − 6 = **9**
6. 12 + 3 = **15**
7. 15 − 7 = **8**
8. 14 + 1 = **15**
9. 17 − 8 = **9**
10. 11 − 7 = **4**
11. 13 − 4 = **9**
12. 11 + 4 = **15**
13. 12 + 6 = **18**
14. 15 − 9 = **6**
15. 11 − 5 = **6**
16. 10 − 1 = **9**
17. 14 + 5 = **19**
18. 16 − 8 = **8**
19. 10 − 2 = **8**
20. 13 − 5 = **8**
21. 10 + 8 = **18**
22. 11 + 2 = **13**
23. 10 + 9 = **19**
24. 11 − 8 = **3**
25. 13 − 8 = **5**
26. 17 − 9 = **8**
27. 16 + 2 = **18**
28. 16 + 1 = **17**
29. 18 − 0 = **18**
30. 15 + 5 = **20**

 21

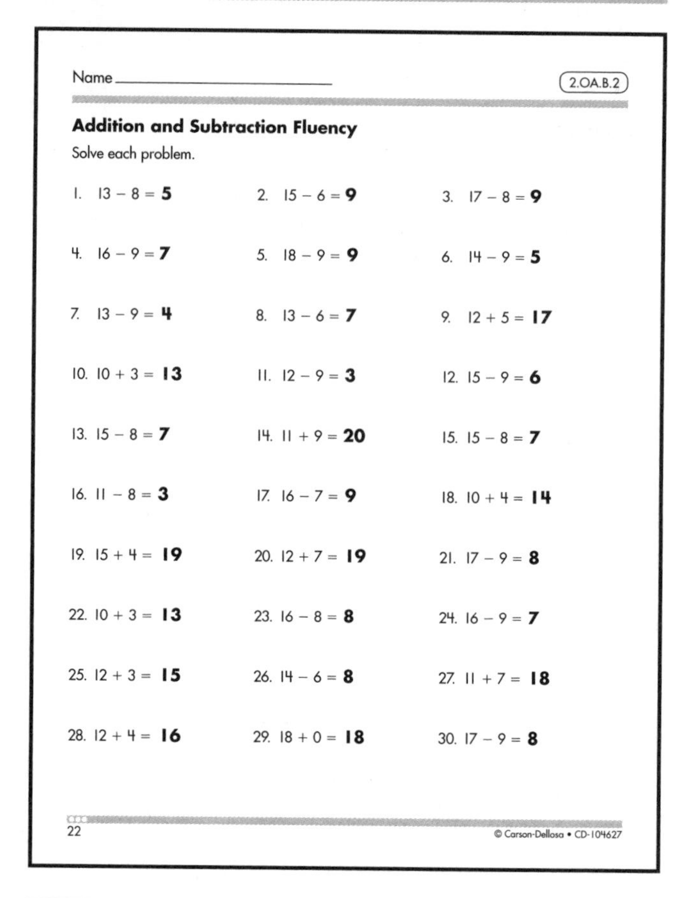
Name ______________________ (2.OA.B.2)

Addition and Subtraction Fluency

Solve each problem.

1. 13 − 8 = **5**
2. 15 − 6 = **9**
3. 17 − 8 = **9**
4. 16 − 9 = **7**
5. 18 − 9 = **9**
6. 14 − 9 = **5**
7. 13 − 9 = **4**
8. 13 − 6 = **7**
9. 12 + 5 = **17**
10. 10 + 3 = **13**
11. 12 − 9 = **3**
12. 15 − 9 = **6**
13. 15 − 8 = **7**
14. 11 + 9 = **20**
15. 15 − 8 = **7**
16. 11 − 8 = **3**
17. 16 − 7 = **9**
18. 10 + 4 = **14**
19. 15 + 4 = **19**
20. 12 + 7 = **19**
21. 17 − 9 = **8**
22. 10 + 3 = **13**
23. 16 − 8 = **8**
24. 16 − 9 = **7**
25. 12 + 3 = **15**
26. 14 − 6 = **8**
27. 11 + 7 = **18**
28. 12 + 4 = **16**
29. 18 + 0 = **18**
30. 17 − 9 = **8**

22

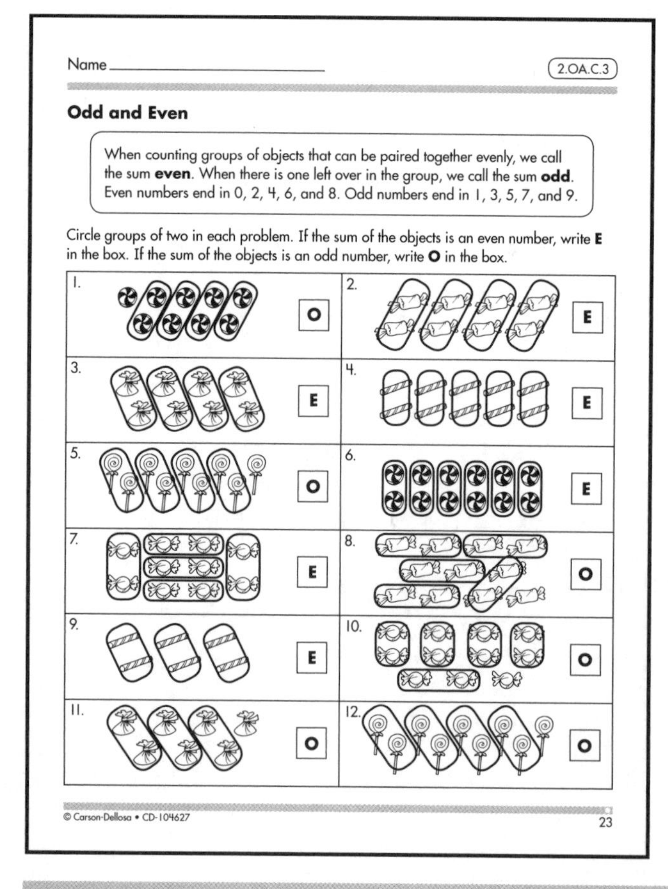
Name ______________________ (2.OA.C.3)

Odd and Even

When counting groups of objects that can be paired together evenly, we call the sum **even**. When there is one left over in the group, we call the sum **odd**. Even numbers end in 0, 2, 4, 6, and 8. Odd numbers end in 1, 3, 5, 7, and 9.

Circle groups of two in each problem. If the sum of the objects is an even number, write **E** in the box. If the sum of the objects is an odd number, write **O** in the box.

1. O
2. E
3. E
4. E
5. O
6. E
7. E
8. O
9. E
10. O
11. O
12. O

 23

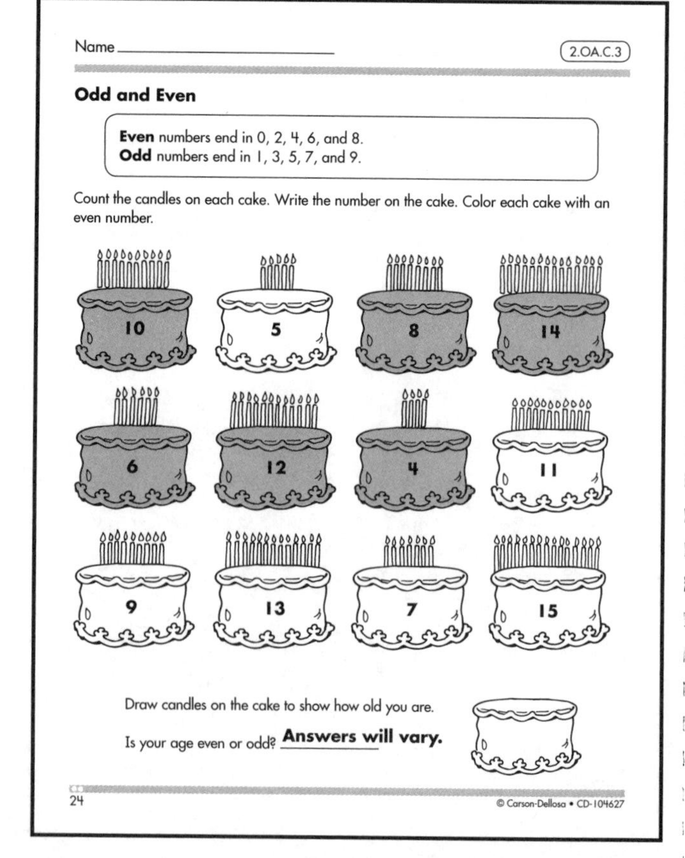
Name ______________________ (2.OA.C.3)

Odd and Even

Even numbers end in 0, 2, 4, 6, and 8.
Odd numbers end in 1, 3, 5, 7, and 9.

Count the candles on each cake. Write the number on the cake. Color each cake with an even number.

10 5 8 14
6 12 4 11
9 13 7 15

Draw candles on the cake to show how old you are.

Is your age even or odd? **Answers will vary.**

24

Answer Key

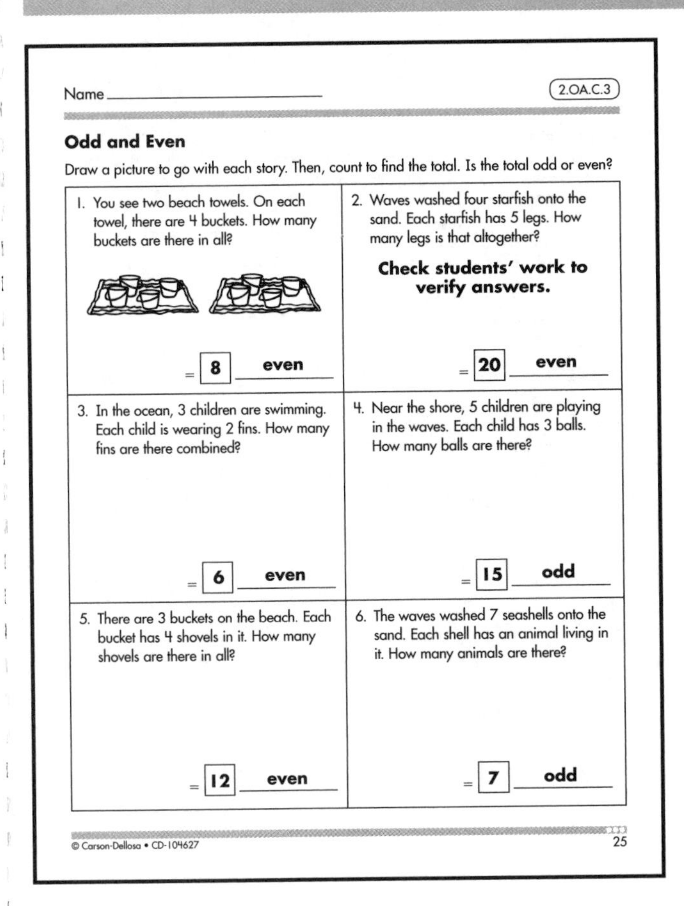

Name ________________ (2.OA.C.3)

Odd and Even

Draw a picture to go with each story. Then, count to find the total. Is the total odd or even?

1. You see two beach towels. On each towel, there are 4 buckets. How many buckets are there in all? = **8** **even**	2. Waves washed four starfish onto the sand. Each starfish has 5 legs. How many legs is that altogether? **Check students' work to verify answers.** = **20** **even**
3. In the ocean, 3 children are swimming. Each child is wearing 2 fins. How many fins are there combined? = **6** **even**	4. Near the shore, 5 children are playing in the waves. Each child has 3 balls. How many balls are there? = **15** **odd**
5. There are 3 buckets on the beach. Each bucket has 4 shovels in it. How many shovels are there in all? = **12** **even**	6. The waves washed 7 seashells onto the sand. Each shell has an animal living in it. How many animals are there? = **7** **odd**

© Carson-Dellosa • CD-104627 25

Name ________________ (2.OA.C.4)

Using Arrays to Add Equal Groups

Objects placed in equal rows and columns are called **arrays**. A row goes across and a column goes down. Use repeated addition to find the sum of the objects in an array.

☆☆☆☆
☆☆☆☆ 2 + 2 + 2 + 2 = **8**

Add the objects in each column to find the total.

1. 2 + 2 + 2 = **6**	2. 5 + 5 + 5 + 5 = **20**
3. **1** + **1** + **1** + **1** + **1** + **1** = **6**	4. **3** + **3** + **3** + **3** + **3** = **15**
5. **4** + **4** + **4** = **12**	6. **3** + **3** + **3** + **3** = **12**

26 © Carson-Dellosa • CD-104627

Name ________________ (2.OA.C.4)

Using Arrays to Add Equal Groups

Add the objects in each column to find the total. Write a number sentence for each array.

1. **3 + 3 + 3 = 9**	2. **3 + 3 + 3 + 3 = 12**
3. **3 + 3 = 6**	4. **3 + 3 + 3 + 3 + 3 = 15**
5. **1 + 1 + 1 + 1 + 1 = 5**	6. **4 + 4 + 4 + 4 + 4 = 20**
7. **2 + 2 = 4**	8. **5 + 5 + 5 = 15**
9. **5 + 5 + 5 + 5 + 5 = 25**	10. **4 + 4 + 4 + 4 = 16**

© Carson-Dellosa • CD-104627 27

Name ________________ (2.OA.C.4)

Using Arrays To Add Equal Groups

Add the columns to find the total number of squares in each rectangle. Write a multiplication sentence for each array.

1. **2+2+2+2+2=10** **2 × 5 = 10**	2. **3+3+3+3+3=15** **3 × 5 = 15**	3. **6 + 6 = 12** **6 × 2 = 12**	4. **4 + 4 + 4 = 12** **4 × 3 = 12**
5. **5+5+5+5+5=25** **5 × 5 = 25**	6. **5 + 5 = 10** **5 × 2 = 10**	7. **4+4+4+4=16** **4 × 4 = 16**	8. **3 + 3 = 6** **3 × 2 = 6**

28 © Carson-Dellosa • CD-104627

Answer Key

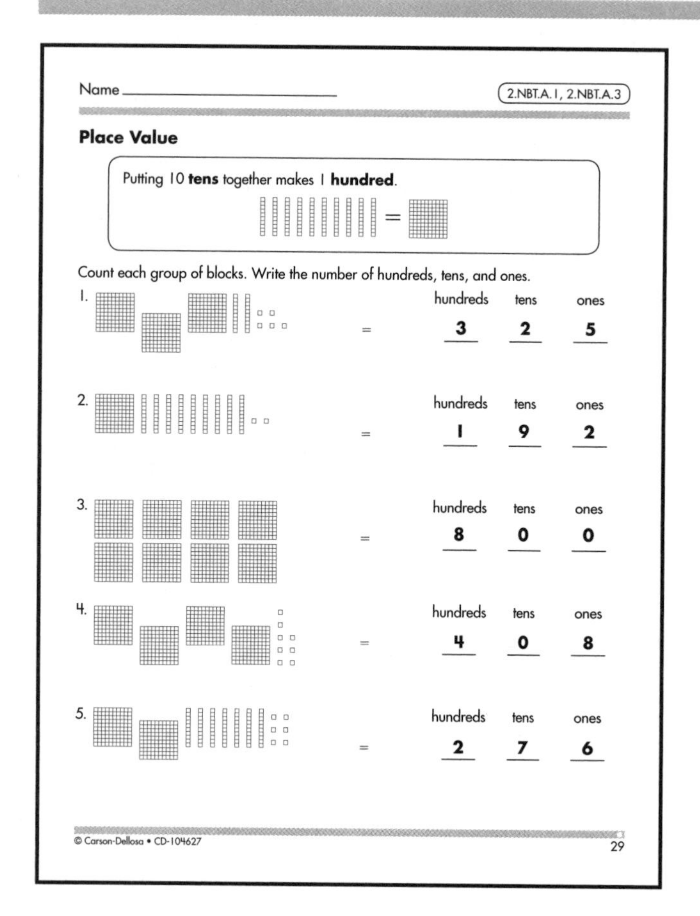

Name ____________________ (2.NBT.A.1, 2.NBT.A.3)

Place Value

Putting 10 **tens** together makes 1 **hundred**.

Count each group of blocks. Write the number of hundreds, tens, and ones.

	hundreds	tens	ones
1.	**3**	**2**	**5**
2.	**1**	**9**	**2**
3.	**8**	**0**	**0**
4.	**4**	**0**	**8**
5.	**2**	**7**	**6**

© Carson-Dellosa • CD-104627 29

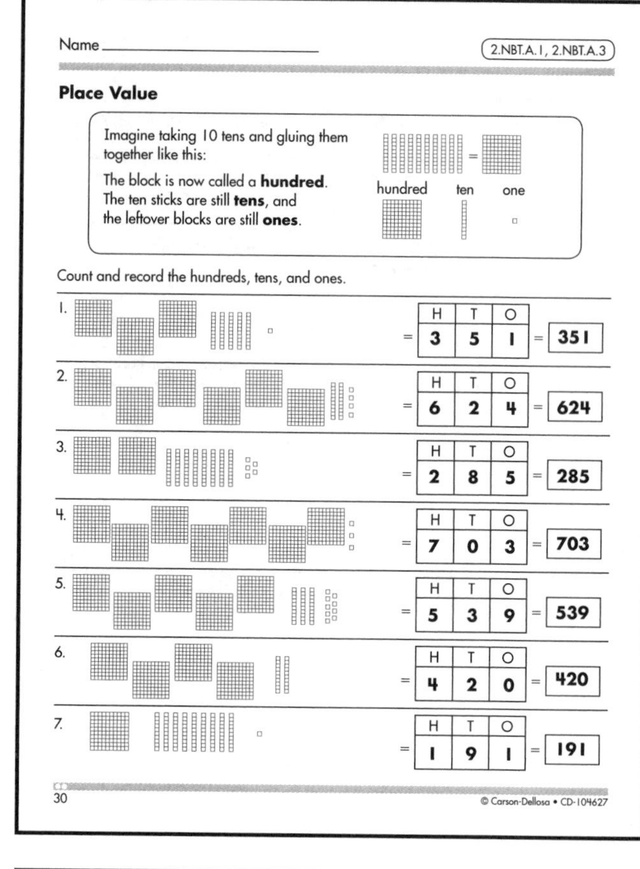

Name ____________________ (2.NBT.A.1, 2.NBT.A.3)

Place Value

Imagine taking 10 tens and gluing them together like this:

The block is now called a **hundred**.
The ten sticks are still **tens**, and the leftover blocks are still **ones**.

hundred ten one

Count and record the hundreds, tens, and ones.

	H	T	O	=
1.	**3**	**5**	**1**	**351**
2.	**6**	**2**	**4**	**624**
3.	**2**	**8**	**5**	**285**
4.	**7**	**0**	**3**	**703**
5.	**5**	**3**	**9**	**539**
6.	**4**	**2**	**0**	**420**
7.	**1**	**9**	**1**	**191**

30 © Carson-Dellosa • CD-104627

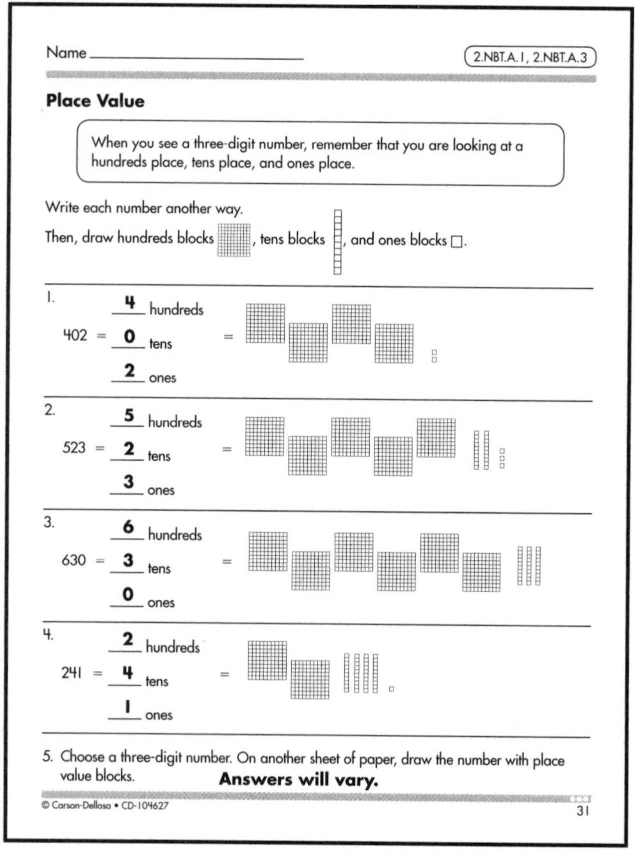

Name ____________________ (2.NBT.A.1, 2.NBT.A.3)

Place Value

When you see a three-digit number, remember that you are looking at a hundreds place, tens place, and ones place.

Write each number another way.
Then, draw hundreds blocks, tens blocks, and ones blocks.

1. 402 = **4** hundreds **0** tens **2** ones =
2. 523 = **5** hundreds **2** tens **3** ones =
3. 630 = **6** hundreds **3** tens **0** ones =
4. 241 = **2** hundreds **4** tens **1** ones =

5. Choose a three-digit number. On another sheet of paper, draw the number with place value blocks. **Answers will vary.**

© Carson-Dellosa • CD-104627 31

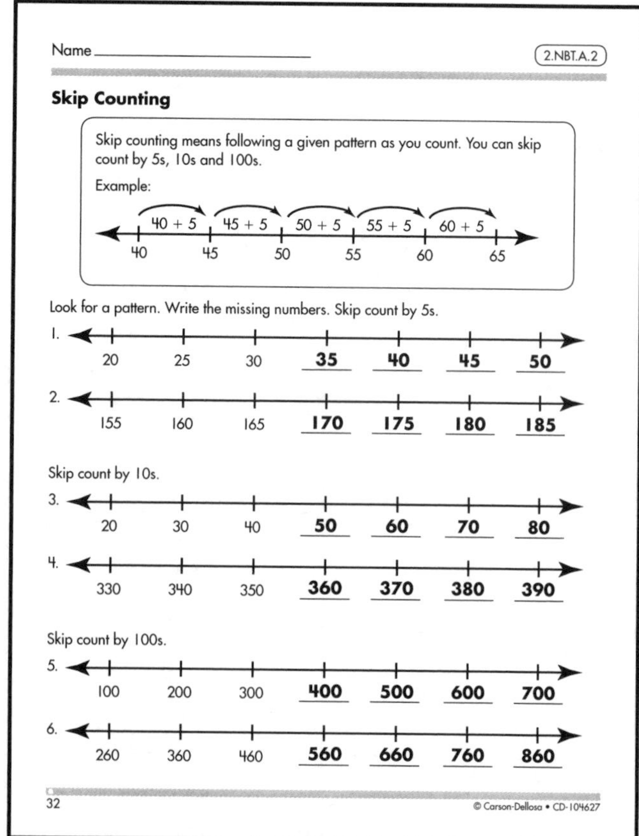

Name ____________________ (2.NBT.A.2)

Skip Counting

Skip counting means following a given pattern as you count. You can skip count by 5s, 10s and 100s.

Example:

40 + 5, 45 + 5, 50 + 5, 55 + 5, 60 + 5

40 45 50 55 60 65

Look for a pattern. Write the missing numbers. Skip count by 5s.

1. 20 25 30 **35** **40** **45** **50**
2. 155 160 165 **170** **175** **180** **185**

Skip count by 10s.

3. 20 30 40 **50** **60** **70** **80**
4. 330 340 350 **360** **370** **380** **390**

Skip count by 100s.

5. 100 200 300 **400** **500** **600** **700**
6. 260 360 460 **560** **660** **760** **860**

32 © Carson-Dellosa • CD-104627

Answer Key

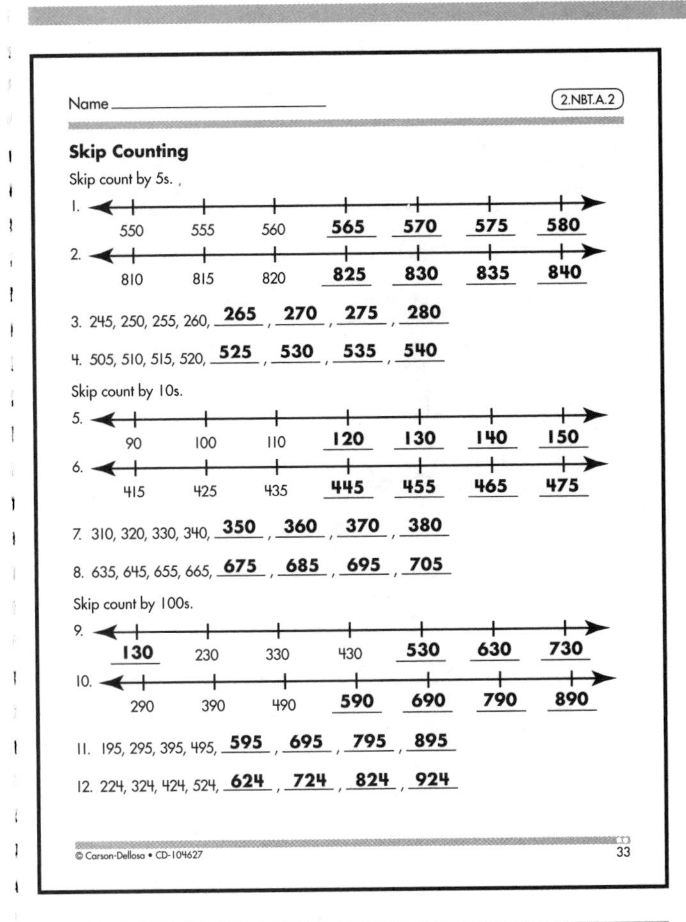
Name ____________ (2.NBT.A.2)

Skip Counting

Skip count by 5s.

1. 550, 555, 560, **565**, **570**, **575**, **580**
2. 810, 815, 820, **825**, **830**, **835**, **840**
3. 245, 250, 255, 260, **265**, **270**, **275**, **280**
4. 505, 510, 515, 520, **525**, **530**, **535**, **540**

Skip count by 10s.

5. 90, 100, 110, **120**, **130**, **140**, **150**
6. 415, 425, 435, **445**, **455**, **465**, **475**
7. 310, 320, 330, 340, **350**, **360**, **370**, **380**
8. 635, 645, 655, 665, **675**, **685**, **695**, **705**

Skip count by 100s.

9. **130**, 230, 330, 430, **530**, **630**, **730**
10. 290, 390, 490, **590**, **690**, **790**, **890**
11. 195, 295, 395, 495, **595**, **695**, **795**, **895**
12. 224, 324, 424, 524, **624**, **724**, **824**, **924**

© Carson-Dellosa • CD-104627 33

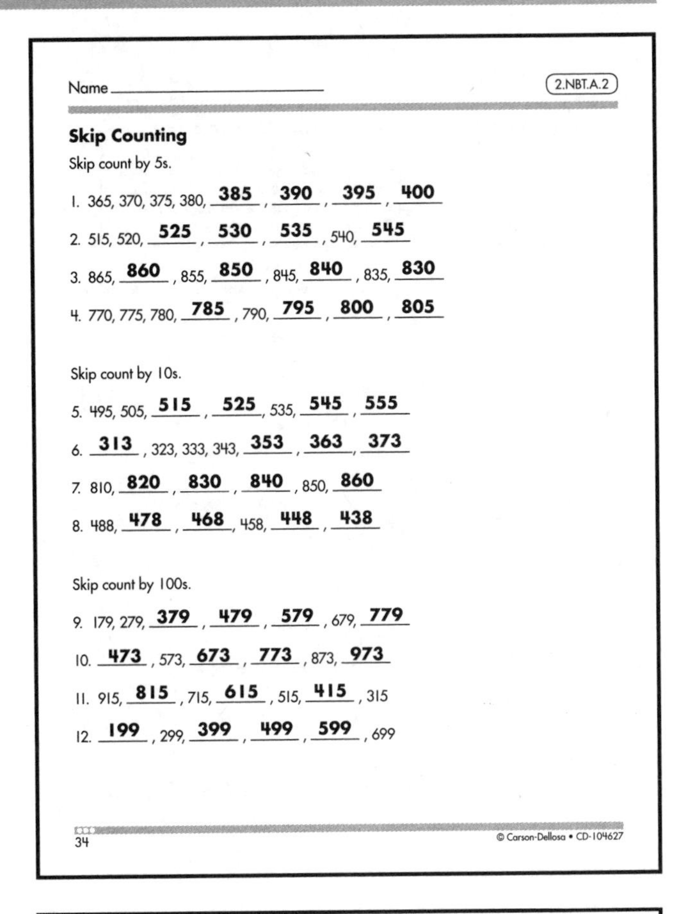
Name ____________ (2.NBT.A.2)

Skip Counting

Skip count by 5s.

1. 365, 370, 375, 380, **385**, **390**, **395**, **400**
2. 515, 520, **525**, **530**, **535**, 540, **545**
3. 865, **860**, 855, **850**, 845, **840**, 835, **830**
4. 770, 775, 780, **785**, 790, **795**, **800**, **805**

Skip count by 10s.

5. 495, 505, **515**, **525**, 535, **545**, **555**
6. **313**, 323, 333, 343, **353**, **363**, **373**
7. 810, **820**, **830**, **840**, 850, **860**
8. 488, **478**, **468**, 458, **448**, **438**

Skip count by 100s.

9. 179, 279, **379**, **479**, **579**, 679, **779**
10. **473**, 573, **673**, **773**, 873, **973**
11. 915, **815**, 715, **615**, 515, **415**, 315
12. **199**, 299, **399**, **499**, **599**, 699

34 © Carson-Dellosa • CD-104627

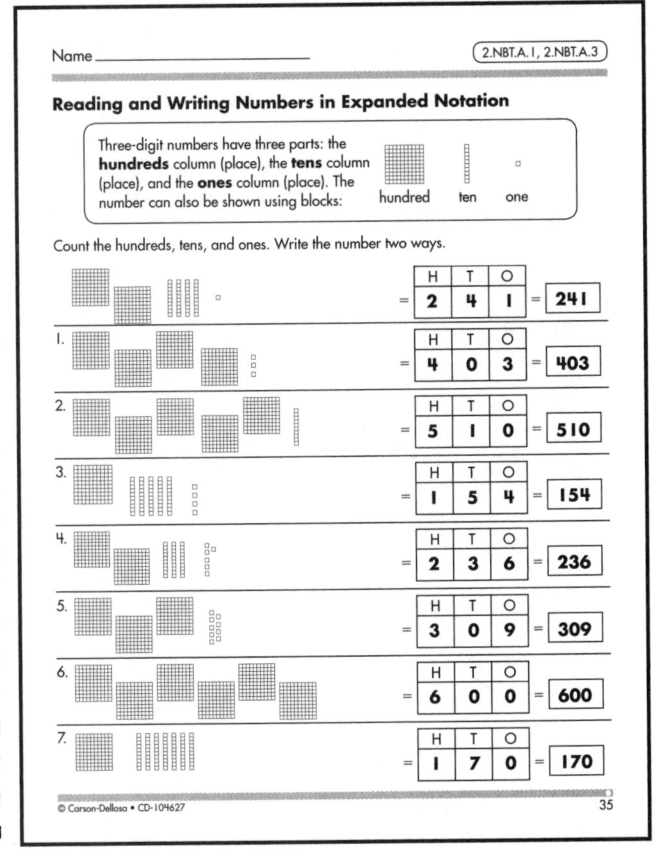
Name ____________ (2.NBT.A.1, 2.NBT.A.3)

Reading and Writing Numbers in Expanded Notation

Three-digit numbers have three parts: the **hundreds** column (place), the **tens** column (place), and the **ones** column (place). The number can also be shown using blocks: hundred ten one

Count the hundreds, tens, and ones. Write the number two ways.

	H	T	O	Number
(example)	**2**	**4**	**1**	**241**
1.	**4**	**0**	**3**	**403**
2.	**5**	**1**	**0**	**510**
3.	**1**	**5**	**4**	**154**
4.	**2**	**3**	**6**	**236**
5.	**3**	**0**	**9**	**309**
6.	**6**	**0**	**0**	**600**
7.	**1**	**7**	**0**	**170**

© Carson-Dellosa • CD-104627 35

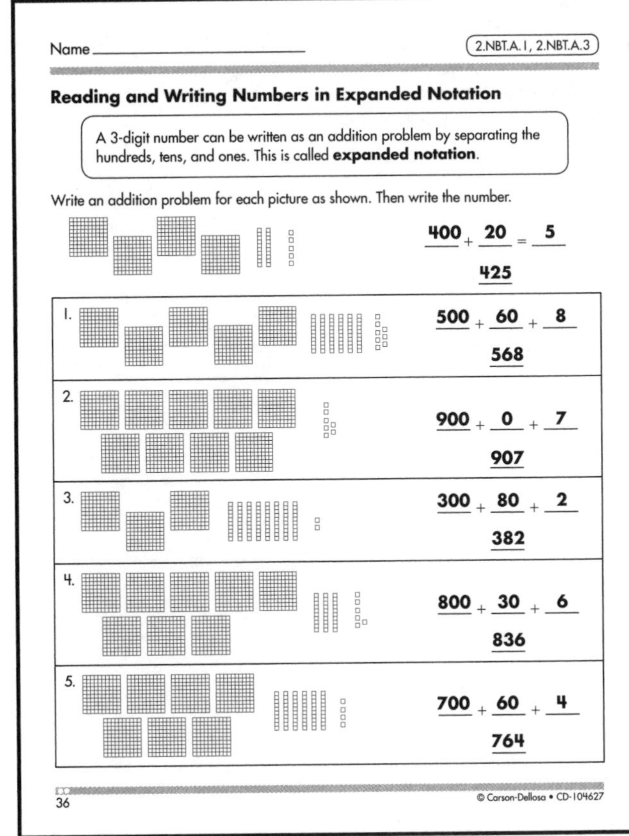
Name ____________ (2.NBT.A.1, 2.NBT.A.3)

Reading and Writing Numbers in Expanded Notation

A 3-digit number can be written as an addition problem by separating the hundreds, tens, and ones. This is called **expanded notation**.

Write an addition problem for each picture as shown. Then write the number.

400 + **20** = **5** — **425**

1. **500** + **60** + **8** — **568**
2. **900** + **0** + **7** — **907**
3. **300** + **80** + **2** — **382**
4. **800** + **30** + **6** — **836**
5. **700** + **60** + **4** — **764**

36 © Carson-Dellosa • CD-104627

Answer Key

Name ____________________ (2.NBT.A.1, 2.NBT.A.3)

Reading and Writing Numbers in Expanded Notation

> To find the correct order of three-digit numbers, always look at the hundreds place first. The number with the most hundreds is the greatest number. If the numbers in the hundreds place are the same, go to the tens place. If they are the same, go to the ones. The higher number may be called greater, greatest or most. The lower number may be called least, less, or lowest.

Write the matching number in each box. Then, sequence the numbers from least to greatest on the lines at the bottom.

1. 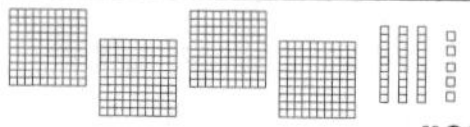 **435**	2. 7 hundreds 6 tens 8 ones **768**
3. nine hundred ten **910**	4. 300 + 70 + 8 **378**
5. 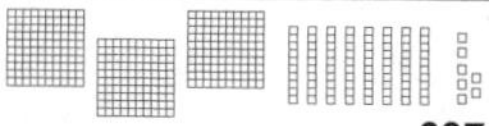 **387**	6. one hundred ninety-nine **199**
7. 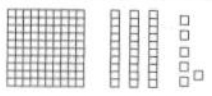 **136**	8. 6 hundreds 8 tens 3 ones **683**
9. 600 + 10 + 4 **614**	10. eight hundred fifteen **815**

Least Greatest

11. 136, **199**, **378**, **387**, **435**, **614**, **683**, **768**, **815**, **910**

Name ____________________ (2.NBT.A.4)

Comparing Numbers

> Comparing numbers means deciding which number is greater and which number is smaller. The symbol > means **greater than** and the symbol < means **less than**. Use these steps to compare:
>
> 1. Are the number of digits the same? If not, the number with the most digits is larger.
> 2. If the number of digits is the same, begin with the digit on the left. Which number has a larger digit? That is the greater number.
> 3. If the digits are the same, move to the next place and find the larger number. If the numbers are the same, they are equal (=).
>
> 671 (<) 2,318 — This number has more digits, so it is greater.
> 564 (>) 372 — 5 is greater.
> 671 (>) 619 — Same, so go to the next digit. 7 is greater than 1.

Write >, <, or = to compare the numbers.

1. 17 (<) 98
2. 298 (<) 300
3. 82 (>) 18
4. 761 (>) 760
5. 79 (>) 23
6. 395 (>) 217
7. 76 (=) 76
8. 514 (=) 514
9. 63 (>) 50
10. 330 (<) 630
11. 102 (=) 102
12. 129 (<) 130

Name ____________________ (2.NBT.A.4)

Comparing Numbers

Write >, <, or = to compare the numbers.

1. 886 (>) 542
2. 130 (>) 13
3. 119 (>) 109
4. 903 (>) 309
5. 984 (=) 984
6. 153 (=) 153
7. 578 (<) 587
8. 600 (>) 300
9. 907 (>) 709
10. 999 (>) 799
11. 534 (<) 990
12. 865 (>) 568
13. 760 (=) 760
14. 712 (>) 233
15. 521 (>) 125
16. 222 (>) 122
17. 18 (<) 189
18. 905 (=) 905
19. Explain how you know that 980 is greater than 880.

Answers will vary.

Name ____________________ (2.NBT.A.4, 2.MD.C.8)

Comparing Numbers

Find the value of each group of coins. Compare the values and write >, < or = in the circle.

1. **40** ¢ 5¢ 5¢ 10¢ 10¢ 10¢ (>) 25¢ **25** ¢
2. **50** ¢ 25¢ 25¢ (=) 10¢ 10¢ 10¢ 10¢ 10¢ **50** ¢
3. **20** ¢ 1¢ 1¢ 1¢ 1¢ 1¢ 5¢ 10¢ (<) 25¢ **25** ¢
4. **30** ¢ 25¢ 5¢ (>) 10¢ 10¢ 5¢ **25** ¢
5. **80** ¢ 10¢ 10¢ 10¢ 10¢ 10¢ 10¢ 10¢ 10¢ (>) 25¢ 25¢ 25¢ **75** ¢

Draw your own coin combinations to make each symbol true.

6. **Answers will vary.**
<

7.
>

8.
=

Answer Key

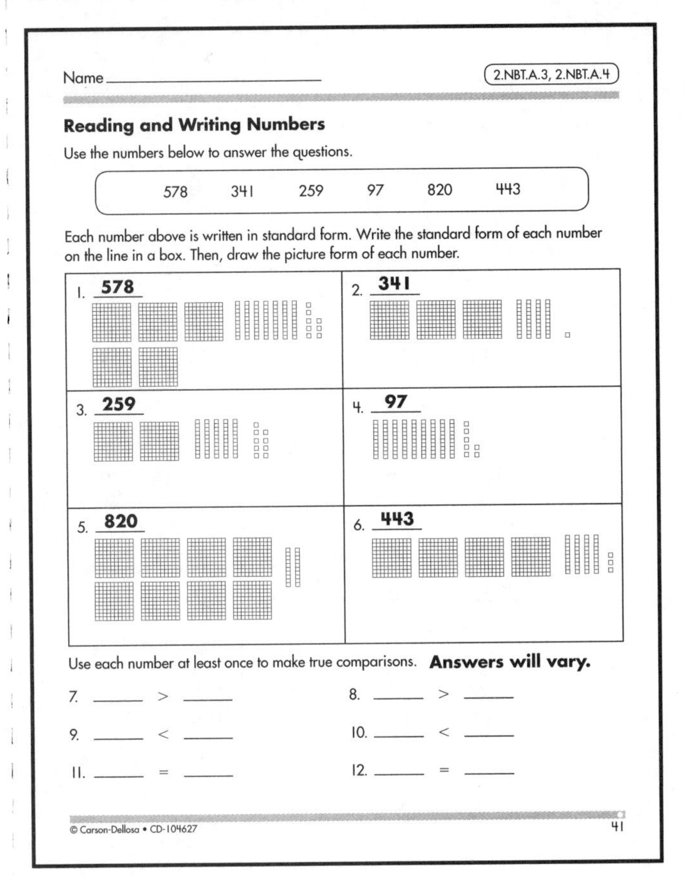

Name ____________________ (2.NBT.A.3, 2.NBT.A.4)

Reading and Writing Numbers

Use the numbers below to answer the questions.

578 341 259 97 820 443

Each number above is written in standard form. Write the standard form of each number on the line in a box. Then, draw the picture form of each number.

1. **578**
2. **341**
3. **259**
4. **97**
5. **820**
6. **443**

Use each number at least once to make true comparisons. **Answers will vary.**

7. ____ > ____
8. ____ > ____
9. ____ < ____
10. ____ < ____
11. ____ = ____
12. ____ = ____

Name ____________________ (2.NBT.A.3)

Reading and Writing Numbers

Write the next four numbers in each pattern. Circle the last one.
Write the expanded form and draw the representational form of the number you circled.
Use these symbols to draw each number: □ = 100 | = 10 ▫ = one

1. 976, 975, 974, 973, **972**, **971**, **(970)**
   Expanded form: **900 + 70 + 0** Representational form:
2. 100, 200, 300, **400**, **500**, **(600)**
   Expanded form: **600 + 0 + 0** Representational form:
3. 367, 377, 387, 397, **407**, **417**, **(427)**
   Expanded form: **400 + 20 + 7** Representational form:
4. 802, 702, 602, 502, **402**, **302**, **(202)**
   Expanded form: **200 + 0 + 2** Representational form:
5. 255, 245, 235, 225, **215**, **205**, **(195)**
   Expanded form: **100 + 90 + 5** Representational form:
6. 338, 348, 358, 368, **378**, **388**, **(398)**
   Expanded form: **300 + 90 + 8** Representational form:
7. 104, 204, 304, 404, **504**, **604**, **(704)**
   Expanded form: **700 + 0 + 4** Representational form:

Name ____________________ (2.NBT.A.3)

Reading and Writing Numbers

Look at the place value blocks. Write the standard, expanded, and word forms of each number.

1. Standard form: **362** Expanded form: **300 + 60 + 2**
   Short word form: **3** hundreds, **6** tens, **2** ones
   Word form: **three hundred sixty-two**
2. Standard form: **238** Expanded form: **200 + 30 + 8**
   Short word form: **2** hundreds, **3** tens, **8** ones
   Word form: **two hundred thirty-eight**
3. Standard form: **54** Expanded form: **0 + 50 + 4**
   Short word form: **0** hundreds, **5** tens, **4** ones
   Word form: **fifty-four**
4. Standard form: **729** Expanded form: **700 + 20 + 9**
   Short word form: **7** hundreds, **2** tens, **9** ones
   Word form: **seven hundred twenty-nine**
5. Standard form: **507** Expanded form: **500 + 7**
   Short word form: **5** hundreds, **0** tens, **7** ones
   Word form: **five hundred seven**

Name ____________________ (2.NBT.B.5)

Adding and Subtracting within 100

Solve each problem.

1. 77 − 3 = **74**
2. 17 + 0 = **17**
3. 44 + 2 = **46**
4. 76 + 3 = **79**
5. 86 + 4 = **90**
6. 42 + 4 = **46**
7. 27 + 1 = **28**
8. 55 + 2 = **57**
9. 77 − 6 = **71**
10. 47 − 6 = **41**
11. 24 + 4 = **28**
12. 78 + 1 = **79**
13. 95 − 0 = **95**
14. 85 − 0 = **85**
15. 75 − 0 = **75**
16. 72 + 1 = **71**
17. 62 + 6 = **68**
18. 57 − 6 = **51**
19. 58 − 8 = **50**
20. 36 − 5 = **31**
21. 61 + 3 = **64**
22. 43 + 3 = **46**
23. 83 − 0 = **83**
24. 89 − 1 = **88**

Answer Key

Name ____________________ (2.NBT.B.5)

Adding and Subtracting within 100

Solve each problem.

1. 74 − 30 **44**	2. 76 + 22 **98**	3. 72 + 23 **95**	4. 74 + 12 **86**	5. 85 − 61 **24**	6. 45 + 24 **69**
7. 60 + 34 **94**	8. 78 + 21 **99**	9. 76 − 26 **50**	10. 43 − 23 **20**	11. 78 + 21 **99**	12. 43 + 53 **96**
13. 16 − 12 **4**	14. 54 − 32 **22**	15. 82 − 42 **40**	16. 33 + 33 **66**	17. 75 + 24 **99**	18. 64 − 23 **41**
19. 45 − 21 **24**	20. 76 − 25 **51**	21. 54 + 45 **99**	22. 67 + 22 **89**	23. 66 − 51 **15**	24. 83 − 62 **21**
25. 52 − 31 **21**	26. 34 + 22 **56**	27. 43 + 34 **77**	28. 80 − 20 **60**	29. 66 + 10 **76**	30. 42 + 42 **84**

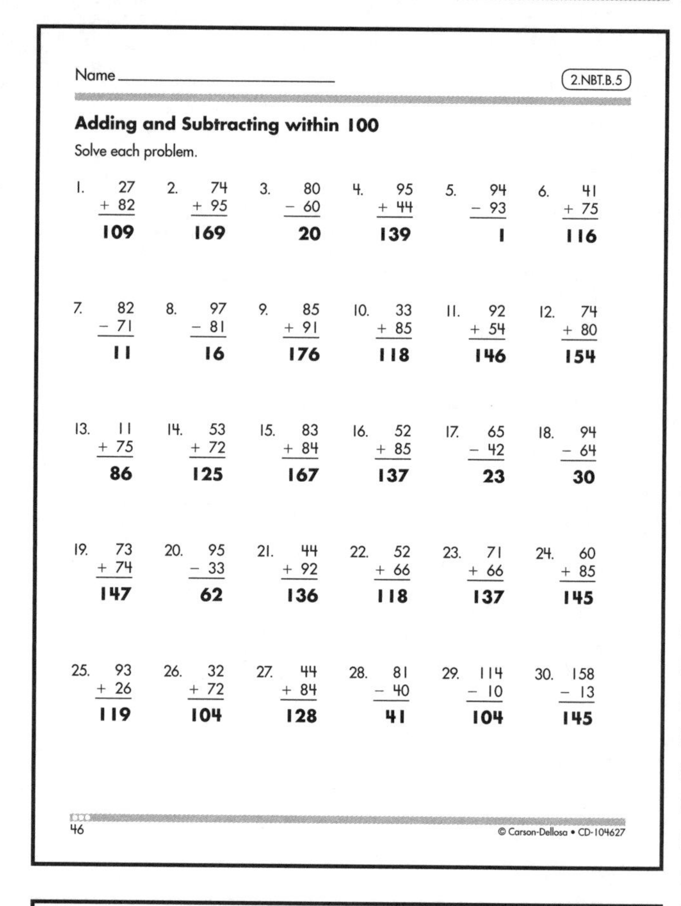

Name ____________________ (2.NBT.B.5)

Adding and Subtracting within 100

Solve each problem.

1. 27 + 82 **109**	2. 74 + 95 **169**	3. 80 − 60 **20**	4. 95 + 44 **139**	5. 94 − 93 **1**	6. 41 + 75 **116**
7. 82 − 71 **11**	8. 97 − 81 **16**	9. 85 + 91 **176**	10. 33 + 85 **118**	11. 92 + 54 **146**	12. 74 + 80 **154**
13. 11 + 75 **86**	14. 53 + 72 **125**	15. 83 + 84 **167**	16. 52 + 85 **137**	17. 65 − 42 **23**	18. 94 − 64 **30**
19. 73 + 74 **147**	20. 95 − 33 **62**	21. 44 + 92 **136**	22. 52 + 66 **118**	23. 71 + 66 **137**	24. 60 + 85 **145**
25. 93 + 26 **119**	26. 32 + 72 **104**	27. 44 + 84 **128**	28. 81 − 40 **41**	29. 114 − 10 **104**	30. 158 − 13 **145**

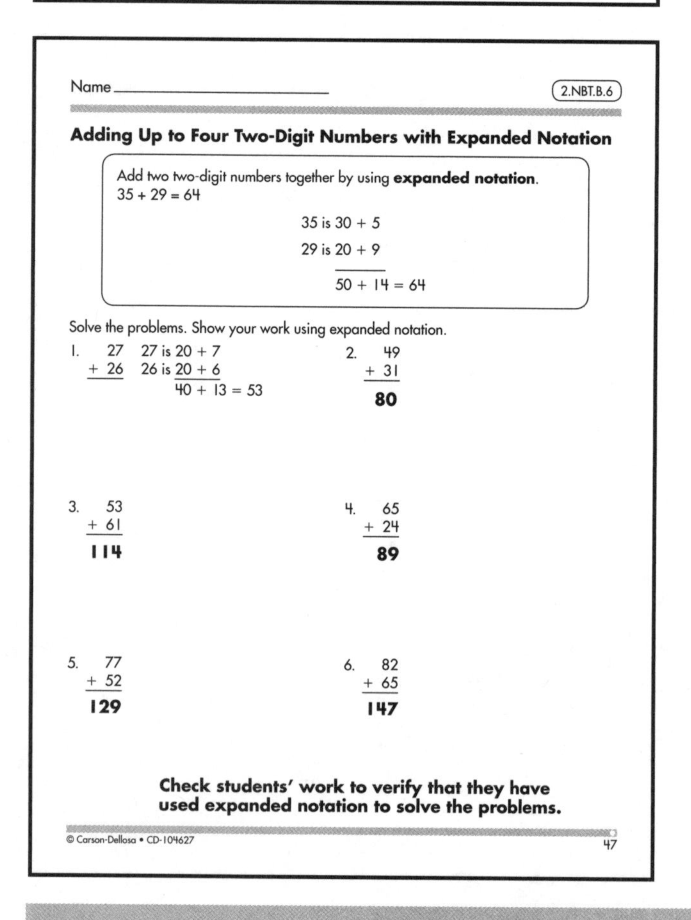

Name ____________________ (2.NBT.B.6)

Adding Up to Four Two-Digit Numbers with Expanded Notation

Add two two-digit numbers together by using **expanded notation**.
35 + 29 = 64

35 is 30 + 5
29 is 20 + 9
50 + 14 = 64

Solve the problems. Show your work using expanded notation.

1. 27 + 26 — 27 is 20 + 7; 26 is 20 + 6; 40 + 13 = 53
2. 49 + 31 **80**
3. 53 + 61 **114**
4. 65 + 24 **89**
5. 77 + 52 **129**
6. 82 + 65 **147**

Check students' work to verify that they have used expanded notation to solve the problems.

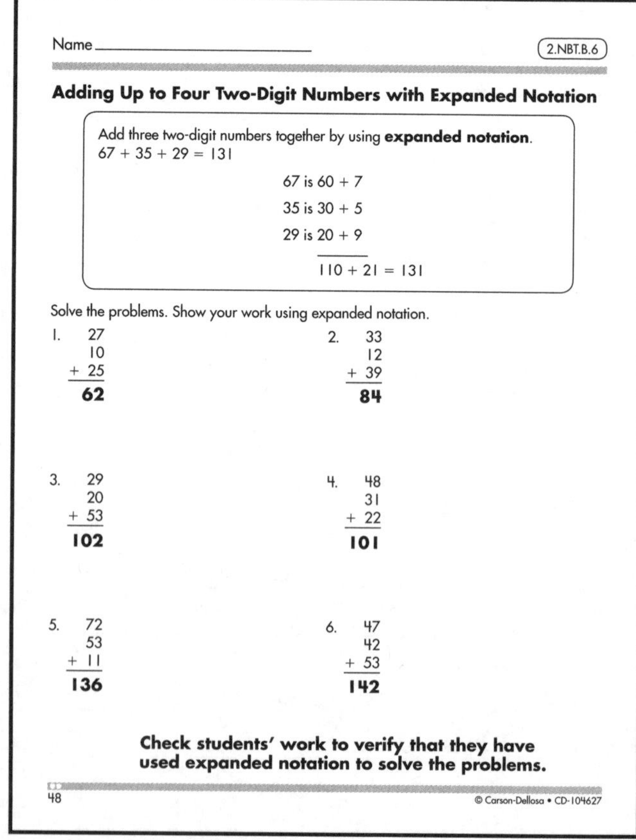

Name ____________________ (2.NBT.B.6)

Adding Up to Four Two-Digit Numbers with Expanded Notation

Add three two-digit numbers together by using **expanded notation**.
67 + 35 + 29 = 131

67 is 60 + 7
35 is 30 + 5
29 is 20 + 9
110 + 21 = 131

Solve the problems. Show your work using expanded notation.

1. 27 + 10 + 25 **62**
2. 33 + 12 + 39 **84**
3. 29 + 20 + 53 **102**
4. 48 + 31 + 22 **101**
5. 72 + 53 + 11 **136**
6. 47 + 42 + 53 **142**

Check students' work to verify that they have used expanded notation to solve the problems.

Answer Key

Name ____________________ (2.NBT.B.6)

Adding Up to Four Two-Digit Numbers with Expanded Notation

Add four two-digit numbers together by using **expanded notation**.
44 + 67 + 35 + 29 = 175

44 is 40 + 4
67 is 60 + 7
35 is 30 + 5
29 is 20 + 9

150 + 25 = 175

Solve the problems. Show your work using expanded notation.

1. 27 + 16 + 64 + 18 = **125**
2. 12 + 25 + 36 + 15 = **88**
3. 21 + 11 + 29 + 42 = **103**
4. 29 + 51 + 53 + 30 = **163**
5. 50 + 24 + 35 + 36 = **145**
6. 12 + 22 + 47 + 28 = **109**

Check students' work to verify that they have used expanded notation to solve the problems.

© Carson-Dellosa • CD-104627 49

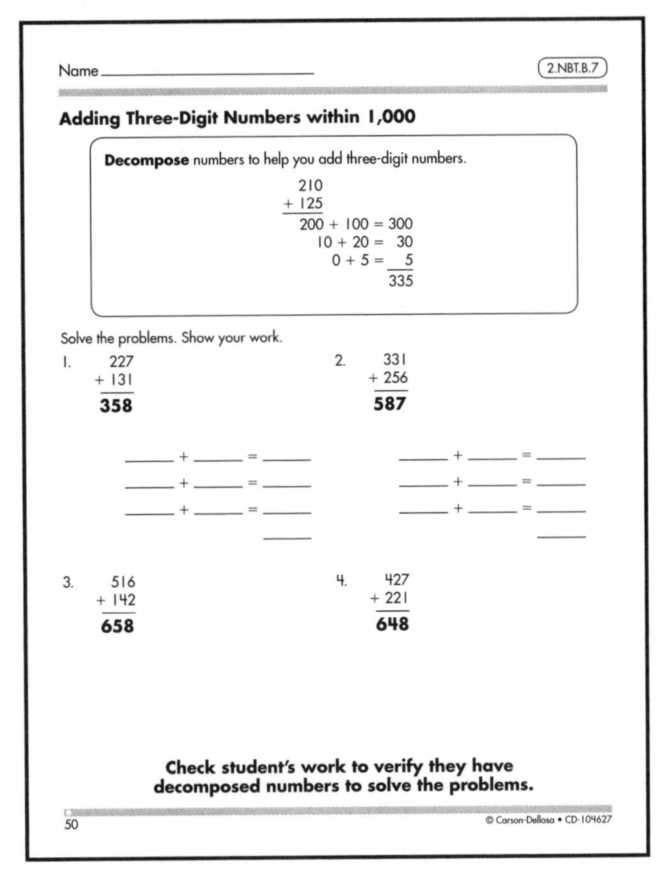

Name ____________________ (2.NBT.B.7)

Adding Three-Digit Numbers within 1,000

Decompose numbers to help you add three-digit numbers.

210
+ 125
200 + 100 = 300
10 + 20 = 30
0 + 5 = 5
335

Solve the problems. Show your work.

1. 227 + 131 = **358**
2. 331 + 256 = **587**

___ + ___ = ___
___ + ___ = ___
___ + ___ = ___

3. 516 + 142 = **658**
4. 427 + 221 = **648**

Check student's work to verify they have decomposed numbers to solve the problems.

50 © Carson-Dellosa • CD-104627

Name ____________________ (2.NBT.B.7)

Adding Three-Digit Numbers within 1,000

Decompose numbers or use other methods, like place value blocks, to help you add three-digit numbers.

261
+ 227
200 + 200 = 400
60 + 20 = 80
1 + 7 = 8
488

400 + 80 + 8 = 488

Solve the problems. Show your work.

1. 155 + 124 = **279**
2. 113 + 371 = **484**
3. 375 + 313 = **688**
4. 277 + 222 = **499**
5. 336 + 132 = **468**
6. 482 + 211 = **693**

Check students' work to verify they have decomposed numbers or used place value blocks to solve the problems.

© Carson-Dellosa • CD-104627 51

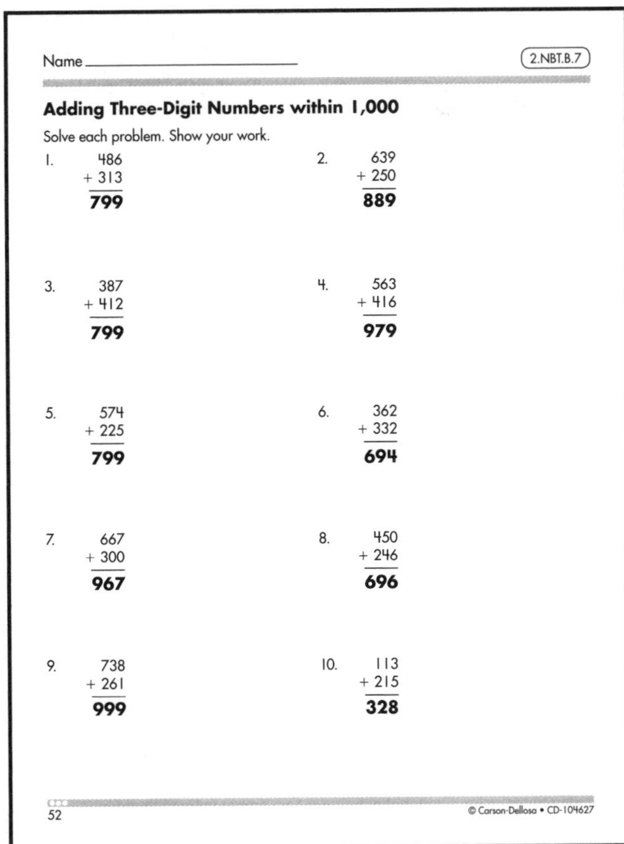

Name ____________________ (2.NBT.B.7)

Adding Three-Digit Numbers within 1,000

Solve each problem. Show your work.

1. 486 + 313 = **799**
2. 639 + 250 = **889**
3. 387 + 412 = **799**
4. 563 + 416 = **979**
5. 574 + 225 = **799**
6. 362 + 332 = **694**
7. 667 + 300 = **967**
8. 450 + 246 = **696**
9. 738 + 261 = **999**
10. 113 + 215 = **328**

52 © Carson-Dellosa • CD-104627

Answer Key

Name ______________________ (2.NBT.B.7)

Subtracting Three-Digit Numbers within 1,000

Decompose numbers to help you subtract three-digit numbers.

```
  289
- 125
  200 - 100 = 100
   80 -  20 =  60
    9 -   5 =   4
              164
```

Solve the problems. Show your work.

1. 577 − 234 = **343**
2. 643 − 521 = **122**

_____ − _____ = _____
_____ − _____ = _____
_____ − _____ = _____
_____

3. 498 − 257 = **241**
4. 873 − 752 = **121**

Check students' work to verify that they have used expanded notation to solve the problems.

© Carson-Dellosa • CD-104627 53

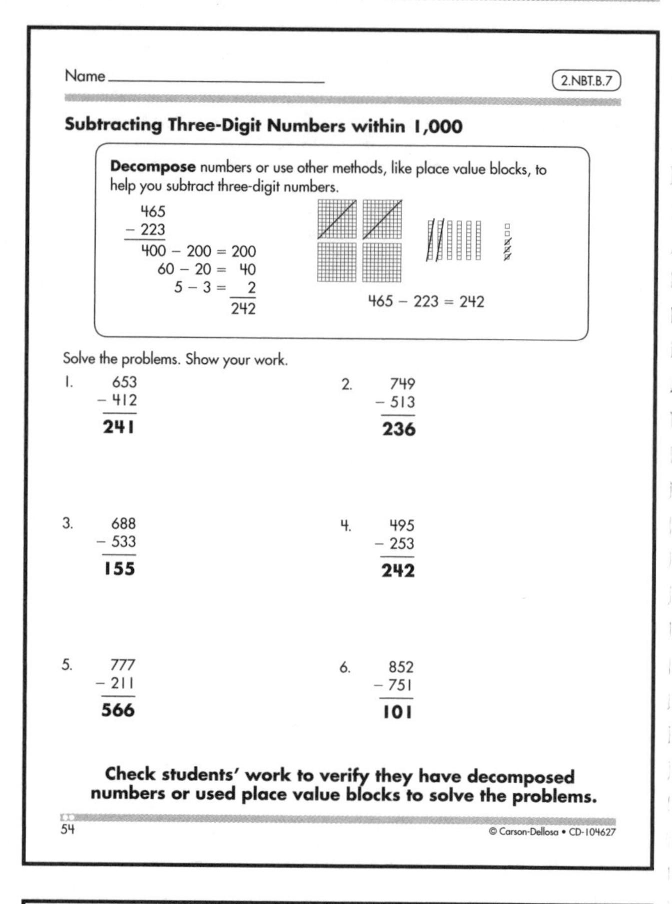

Name ______________________ (2.NBT.B.7)

Subtracting Three-Digit Numbers within 1,000

Decompose numbers or use other methods, like place value blocks, to help you subtract three-digit numbers.

```
  465
- 223
  400 - 200 = 200
   60 -  20 =  40
    5 -   3 =   2
              242
```

465 − 223 = 242

Solve the problems. Show your work.

1. 653 − 412 = **241**
2. 749 − 513 = **236**
3. 688 − 533 = **155**
4. 495 − 253 = **242**
5. 777 − 211 = **566**
6. 852 − 751 = **101**

Check students' work to verify they have decomposed numbers or used place value blocks to solve the problems.

54 © Carson-Dellosa • CD-104627

Name ______________________ (2.NBT.B.7)

Subtracting Three-Digit Numbers within 1,000

Solve each problem. Show your work.

1. 981 − 360 = **621**
2. 213 − 101 = **112**
3. 999 − 222 = **777**
4. 548 − 413 = **135**
5. 179 − 156 = **23**
6. 887 − 544 = **343**
7. 324 − 123 = **201**
8. 495 − 262 = **233**
9. 474 − 333 = **141**
10. 262 − 131 = **131**

© Carson-Dellosa • CD-104627 55

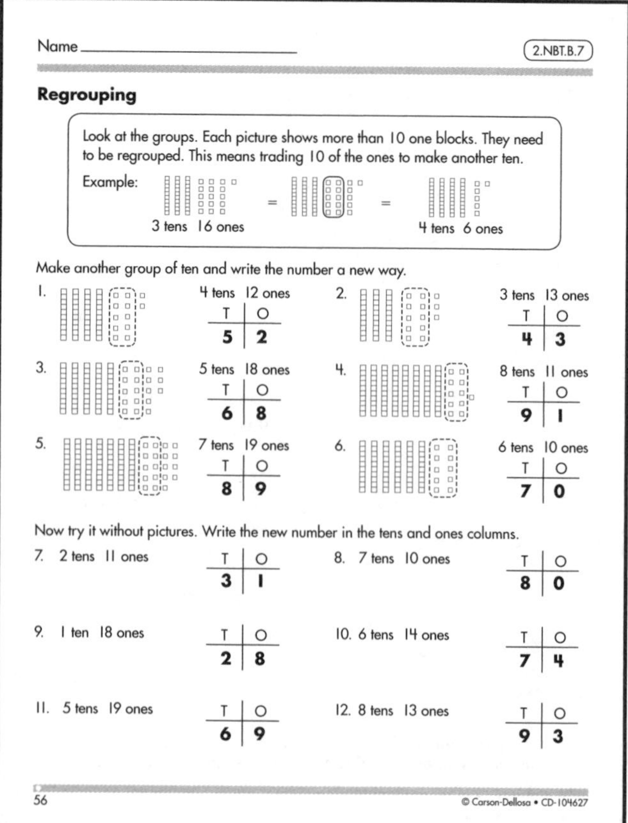

Name ______________________ (2.NBT.B.7)

Regrouping

Look at the groups. Each picture shows more than 10 one blocks. They need to be regrouped. This means trading 10 of the ones to make another ten.

Example: 3 tens 16 ones = 4 tens 6 ones

Make another group of ten and write the number a new way.

Problem	Given	T	O
1.	4 tens 12 ones	**5**	**2**
2.	3 tens 13 ones	**4**	**3**
3.	5 tens 18 ones	**6**	**8**
4.	8 tens 11 ones	**9**	**1**
5.	7 tens 19 ones	**8**	**9**
6.	6 tens 10 ones	**7**	**0**

Now try it without pictures. Write the new number in the tens and ones columns.

Problem	Given	T	O
7.	2 tens 11 ones	**3**	**1**
8.	7 tens 10 ones	**8**	**0**
9.	1 ten 18 ones	**2**	**8**
10.	6 tens 14 ones	**7**	**4**
11.	5 tens 19 ones	**6**	**9**
12.	8 tens 13 ones	**9**	**3**

56 © Carson-Dellosa • CD-104627

Answer Key

Name ______________ (2.NBT.B.7)

Regrouping

Write out each problem in short word form. Add the ones. Add the tens. Regroup. Write the answer in number form.

1. 72 + 83 can be written as:
   7 tens + 2 ones
   \+ 8 tens + 3 ones
   15 tens + **5** ones
   Regroup: **1** hundreds + **5** tens + **5** ones = **155**

2. 54 + 38 can be written as:
   5 tens + 4 ones
   \+ 3 tens + 8 ones
   8 tens + **12** ones
   Regroup: **0** hundreds + **9** tens + **2** ones = **92**

3. 97 + 72 can be written as:
   9 tens + **7** ones
   \+ **7** tens + **2** ones
   16 tens + **9** ones
   Regroup: **1** hundreds + **6** tens + **9** ones = **169**

4. 57 + 28 can be written as:
   5 tens + **7** ones
   \+ **2** tens + **8** ones
   7 tens + **15** ones
   Regroup: **0** hundreds + **8** tens + **5** ones = **85**

5. 68 + 71 can be written as:
   6 tens + **8** ones
   \+ **7** tens + **1** ones
   13 tens + **9** ones
   Regroup: **1** hundreds + **3** tens + **9** ones = **139**

6. 61 + 29 can be written as:
   6 tens + **1** ones
   \+ **2** tens + **9** ones
   8 tens + **10** ones
   Regroup: **0** hundreds + **9** tens + **0** ones = **90**

© Carson-Dellosa • CD-104627 57

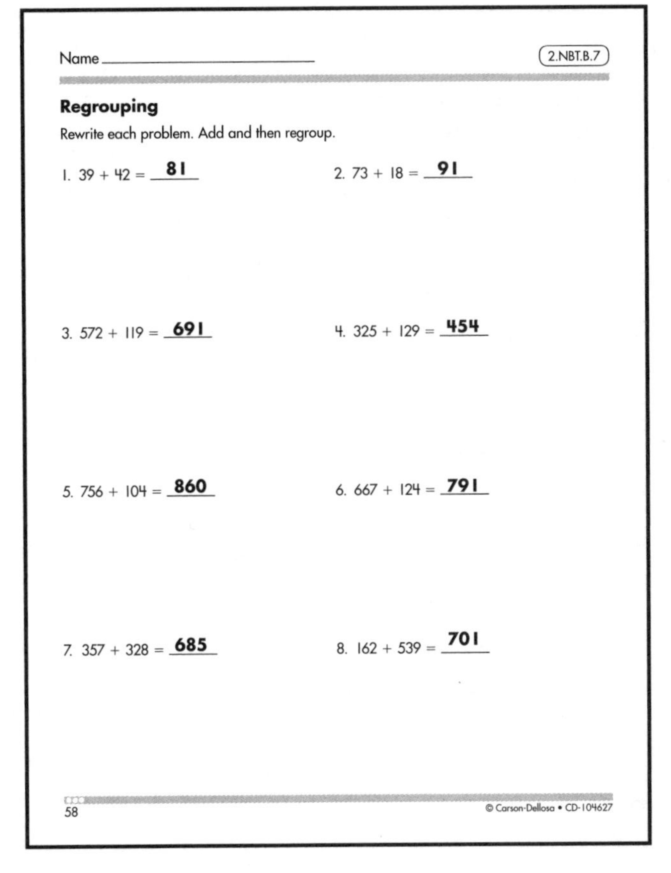

Name ______________ (2.NBT.B.7)

Regrouping

Rewrite each problem. Add and then regroup.

1. 39 + 42 = **81**
2. 73 + 18 = **91**
3. 572 + 119 = **691**
4. 325 + 129 = **454**
5. 756 + 104 = **860**
6. 667 + 124 = **791**
7. 357 + 328 = **685**
8. 162 + 539 = **701**

58 © Carson-Dellosa • CD-104627

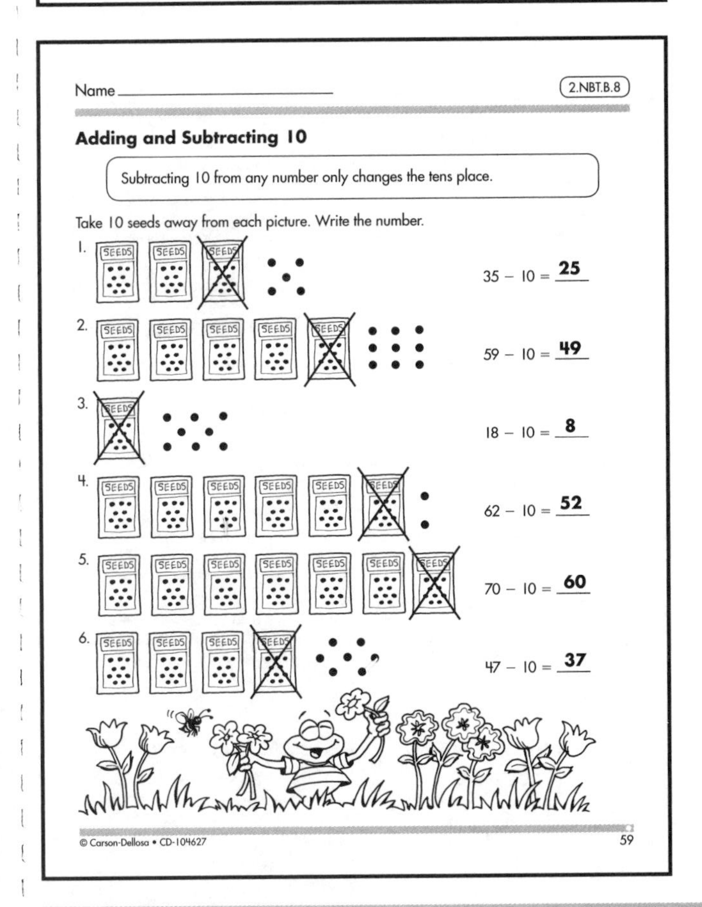

Name ______________ (2.NBT.B.8)

Adding and Subtracting 10

Subtracting 10 from any number only changes the tens place.

Take 10 seeds away from each picture. Write the number.

1. 35 − 10 = **25**
2. 59 − 10 = **49**
3. 18 − 10 = **8**
4. 62 − 10 = **52**
5. 70 − 10 = **60**
6. 47 − 10 = **37**

© Carson-Dellosa • CD-104627 59

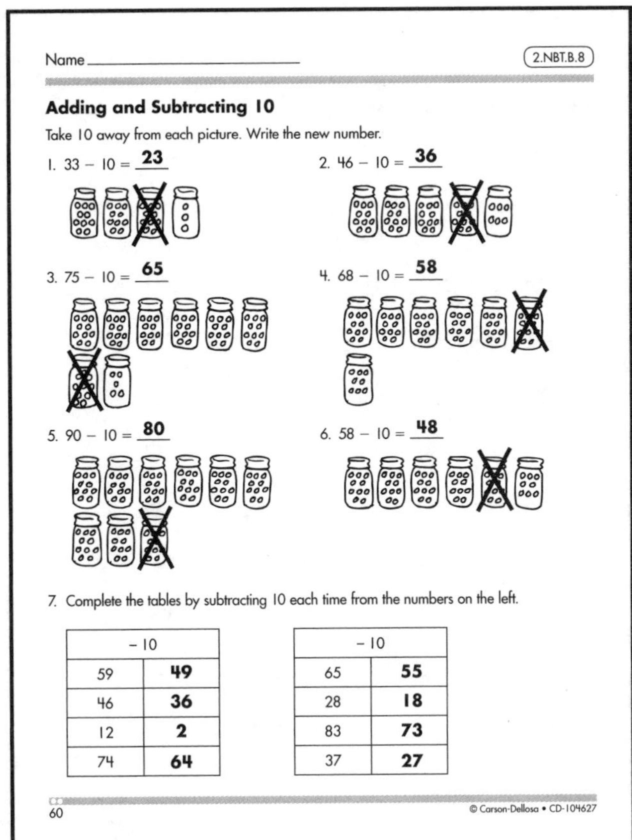

Name ______________ (2.NBT.B.8)

Adding and Subtracting 10

Take 10 away from each picture. Write the new number.

1. 33 − 10 = **23**
2. 46 − 10 = **36**
3. 75 − 10 = **65**
4. 68 − 10 = **58**
5. 90 − 10 = **80**
6. 58 − 10 = **48**
7. Complete the tables by subtracting 10 each time from the numbers on the left.

− 10	
59	**49**
46	**36**
12	**2**
74	**64**

− 10	
65	**55**
28	**18**
83	**73**
37	**27**

60 © Carson-Dellosa • CD-104627

Answer Key

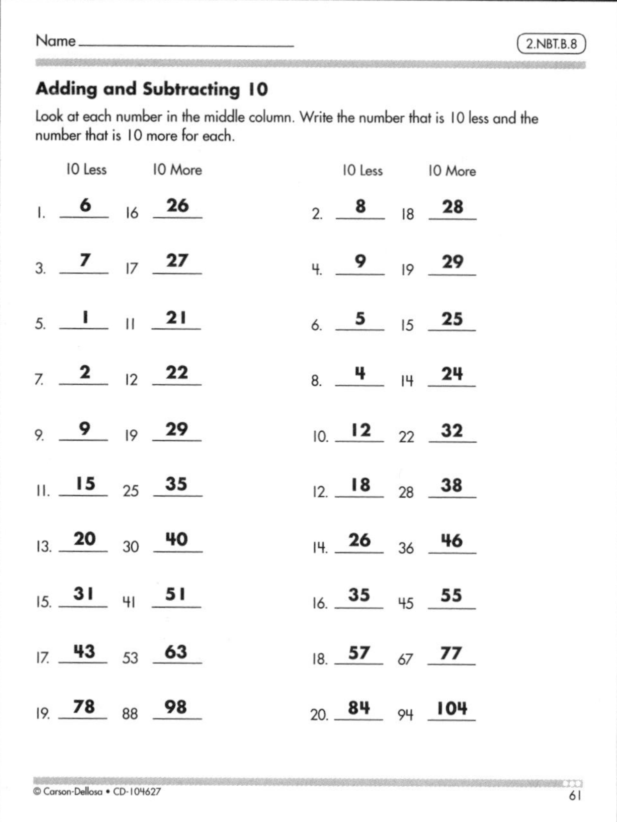

Name ______________________ (2.NBT.B.8)

Adding and Subtracting 10

Look at each number in the middle column. Write the number that is 10 less and the number that is 10 more for each.

	10 Less		10 More		10 Less		10 More
1.	**6**	16	**26**	2.	**8**	18	**28**
3.	**7**	17	**27**	4.	**9**	19	**29**
5.	**1**	11	**21**	6.	**5**	15	**25**
7.	**2**	12	**22**	8.	**4**	14	**24**
9.	**9**	19	**29**	10.	**12**	22	**32**
11.	**15**	25	**35**	12.	**18**	28	**38**
13.	**20**	30	**40**	14.	**26**	36	**46**
15.	**31**	41	**51**	16.	**35**	45	**55**
17.	**43**	53	**63**	18.	**57**	67	**77**
19.	**78**	88	**98**	20.	**84**	94	**104**

© Carson-Dellosa • CD-104627 61

Name ______________________ (2.NBT.B.8)

Adding and Subtracting Hundreds

Adding and subtracting groups of hundreds only changes the hundreds place.

Scientists are always looking at our changing star patterns. Imagine hundredsof new stars being discovered while hundreds of others burn out.

Complete the star chart by adding and subtracting 100.

	Number of ★ Before	+ Discovered ★	= Number of ★ Now
1.	881	+ 100	**981**
2.	763	+ 200	**963**
3.	405	+ 300	**705**
4.	312	+ 400	**712**

	Number of ★ Before	– Burned Out ★	= Number of ★ Now
5.	962	– 200	**762**
6.	814	– 300	**514**
7.	730	– 400	**330**
8.	689	– 500	**189**
9.	978	– 600	**378**

10. Explain how you know 300 – 100 = 200. **Answers will vary.**

62 © Carson-Dellosa • CD-104627

Name ______________________ (2.NBT.B.8)

Adding and Subtracting Hundreds

Look at each number in the middle column. Write the number that is 100 less and the number that is 100 more for each.

	100 Less		100 More		100 Less		100 More
1.	**106**	206	**306**	2.	**699**	799	**899**
3.	**414**	514	**614**	4.	**391**	491	**591**
5.	**207**	307	**407**	6.	**744**	844	**944**
7.	**29**	129	**229**	8.	**371**	471	**571**
9.	**562**	662	**762**	10.	**122**	222	**322**
11.	**125**	225	**325**	12.	**528**	628	**728**
13.	**230**	330	**430**	14.	**263**	363	**463**
15.	**310**	410	**510**	16.	**449**	549	**649**
17.	**413**	513	**613**	18.	**570**	670	**770**
19.	**708**	808	**908**	20.	**800**	900	**1,000**

© Carson-Dellosa • CD-104627 63

Name ______________________ (2.NBT.B.8)

Adding and Subtracting Hundreds

Look for key words that tell you to add or subtract. Solve. Show your work.

1. Fran read 100 pages on Saturday. She read 87 pages on Sunday. How many more pages did Fran read on Saturday?

   100 – 87 = 13
   13 pages

2. Carlos counted 200 books in his classroom. Troy counted 88 books. How many books did they count in all?

   200 + 88 = 288
   288 books

3. Jasmine read for 63 minutes in the morning. After lunch, she read for 100 minutes. Before she went to bed, she read for 10 minutes. How many minutes did Jasmine read altogether?

   63 + 100 + 10 = 173
   173 minutes

4. Molly bought a book with 315 pages. She has read 215 pages. How many pages does she have left?

   315 – 215 = 100
   100 pages

5. Emma counted 397 words in her book. Andy counted 197 words in his book. How many more words are in Emma's book?

   397 – 197 = 200
   200 words

6. Grant has read 200 books from the library. Jason has read 125 books. How many books have they read in all?

   200 + 125 = 325
   325 books

64 © Carson-Dellosa • CD-104627

Answer Key

Name ____________________ (2.NBT.B.8)

Addition and Subtraction with 10 or 100

Solve each problem.

1. 26 − 10 = **16**
2. 37 + 10 = **47**
3. 98 − 10 = **88**
4. 94 − 10 = **84**
5. 66 + 10 = **76**
6. 76 − 10 = **66**
7. 34 + 10 = **44**
8. 52 + 10 = **62**
9. 31 + 10 = **41**
10. 80 + 10 = **90**
11. 86 + 10 = **96**
12. 37 + 10 = **47**
13. 63 + 10 = **73**
14. 80 − 10 = **70**
15. 97 − 10 = **87**
16. 93 − 10 = **83**
17. 83 + 10 = **93**
18. 71 − 10 = **61**

Name ____________________ (2.NBT.B.8)

Addition and Subtraction with 10 or 100

Solve each problem.

1. 81 + 10 = **91**
2. 10 + 60 = **70**
3. 10 + 43 = **53**
4. 63 − 10 = **53**
5. 44 − 10 = **34**
6. 72 + 10 = **82**
7. 61 − 10 = **51**
8. 44 + 10 = **54**
9. 10 + 42 = **52**
10. 19 − 10 = **9**
11. 37 − 10 = **27**
12. 10 + 32 = **42**
13. 10 + 75 = **85**
14. 62 − 10 = **52**
15. 10 + 32 = **42**
16. 10 + 21 = **31**
17. 21 − 10 = **11**
18. 10 + 41 = **51**
19. 10 + 53 = **63**
20. 42 − 10 = **32**
21. 60 − 10 = **50**
22. 10 + 46 = **56**
23. 10 + 61 = **71**
24. 111 − 10 = **101**

Name ____________________ (2.NBT.B.8)

Addition and Subtraction with 10 or 100

Solve each problem.

1. 651 − 10 = **641**
2. 994 − 10 = **984**
3. 896 − 10 = **886**
4. 862 − 10 = **852**
5. 379 − 10 = **369**
6. 977 − 10 = **967**
7. 795 − 100 = **695**
8. 785 − 100 = **685**
9. 996 − 100 = **896**
10. 899 − 100 = **799**
11. 439 − 100 = **339**
12. 769 − 100 = **669**
13. 983 + 10 = **993**
14. 998 + 10 = **1,008**
15. 543 + 10 = **553**
16. 978 + 10 = **988**
17. 398 + 10 = **408**
18. 300 + 10 = **310**
19. 925 + 100 = **1,025**
20. 170 + 100 = **270**
21. 286 + 100 = **386**
22. 487 + 100 = **587**
23. 664 − 100 = **564**
24. 856 + 100 = **956**

25. Explain how to add 10 to a number. **Answers will vary.**

Name ____________________ (2.NBT.A.3, 2.NBT.A.4, 2.NBT.B.9)

Explain Your Reasoning

Explain your reasoning when comparing numbers by looking at the place values of the two numbers.

five tens and five ones > 32

This is true because there are 5 tens in 55 and only 3 tens in 32.

60 + 7 < four tens and three ones

This is false because there are more tens in 67 than in 43.

45 = forty-five

This is true because there are the same amount of tens and ones in each number.

Are these comparisons true or false? Circle **True** or **False**. Explain your reasoning.

1. twenty-nine > 30 + 4 — True / (False circled)
   Why? **Check students' work to verify accurate reasoning.**
2. sixty = 40 + 10 — True / (False circled)
   Why? ________
3. 53 > eight tens and two ones — True / (False circled)
   Why? ________
4. six tens + two ones < 80 + 9 — (True circled) / False
   Why? ________
5. forty-two = 42 — (True circled) / False
   Why? ________

Answer Key

Name ________________ (2.NBT.A.3, 2.NBT.A.4, 2.NBT.B.9)

Explain Your Reasoning

Explain your reasoning when comparing numbers by looking at the place values of the two numbers.

three hundreds + five tens + five ones > 232

This is a true statement because there are 3 hundreds in 355 and only 2 hundreds in 232.

400 + 60 + 7 < 500 + 40 + 3

This is a false statement because there are more hundreds in 543 than in 467.

145 = one hundred forty-five

This is true because there are the same amount of hundreds, tens, and ones in each number.

Are these comparisons true or false? Circle **True** or **False**. Explain your reasoning.

1. 4 hundreds + 4 ones > 9 tens and 6 ones — (True) False
   Why? **Check students' work to verify accurate reasoning.**
2. 2 hundreds + 4 tens + 2 ones < 422 — (True) False
   Why? ________
3. 555 > 6 hundreds — True (False)
   Why? ________
4. 700 + 50 + 8 = Seven hundred fifty-eight — (True) False
   Why? ________
5. 3 hundreds + 7 ones > 370 — True (False)
   Why? ________

Name ________________ (2.NBT.A.3, 2.NBT.A.4, 2.NBT.B.9)

Explain Your Reasoning

Are these comparisons true or false? Circle **True** or **False**. Explain your reasoning.

1. 4 tens + 2 hundreds + 2 ones < 422 — True (False)
   Why? **Check student's work to verify accurate resoning.**
2. 988 = Nine hundred ninety-eight — True (False)
   Why? ________
3. 500 + 30 + 3 < 678 — (True) False
   Why? ________
4. Three hundred + ninety + six > 396 — True (False)
   Why? ________
5. 9 ones + six hundreds + twenty > 200 + 60 + 9 — (True) False
   Why? ________
6. four hundred three < 430 — (True) False
   Why? ________
7. 70 + 700 + 2 > seven hundred eighty-seven — True (False)
   Why? ________
8. 561 > 8 hundreds + 6 ones + 7 tens — (True) False
   Why? ________

Name ________________ (2.MD.A.1, 2.MD.A.3, 2.MD.A.4)

Measuring in Inches and Centimeters

Have you ever noticed that some rulers have numbers on both sides? One side shows **inches** and the other shows **centimeters**. To measure length, place the starting point of your ruler on the beginning of the line. Then read the nearest number at the end of the line.

cm 1 2 3 4 5 6 7 8 9 10 11 12 13 14 15

Measure these lines in inches and centimeters. Record the length.

1. This line is **2** **inches** long.
2. This line is **3** **centimeters** long.
3. This line is **5** **inches** long.
4. This line is **5** **centimeters** long.
5. This line is **6** **inches** long.
6. This line is **4** **inches** long.
7. This line is **12** **centimeters** long.
8. This line is **9** **centimeters** long.
9. How much longer is line 5 than 6? **2** inches
10. How much longer is line 7 than 8? **3** centimeters

Name ________________ (2.MD.A.1, 2.MD.A.3, 2.MD.A.4)

Measuring in Inches and Centimeters

Estimate the length of each line. Measure the lines. Record the actual measurement. Write each length in inches or centimeters.

	Estimate	Actual
1. **Students' estimates will vary.**	___ in.	**6** in.
2.	___ cm	**9** cm
3.	___ in.	**1.5** in.
4.	___ cm	**11** cm
5.	___ cm	**2** cm
6.	___ in.	**2.5** in.
7.	___ cm	**17** cm

Draw a line to match the given measurement.

8. 6 inches
9. 17 centimeters
10. 8 inches

Check student answers to verify accurate measurements.

Answer Key

Name ______________ (2.MD.A.1, 2.MD.A.3, 2.MD.A.4)

Measuring in Inches and Centimeters

Estimate the length of each line. Measure the lines. Record the actual measurement. Write each length in inches or centimeters.

	Estimate	Actual
1.	____ cm	**5** cm
2.	____ in.	**6** in.
3.	____ cm	**2.5** cm
4.	____ in.	**7** in.
5.	____ in.	**3** in.

Students' estimates will vary.

Draw a line to match the given measurement.

6. 16 centimeters
7. 9 centimeters
8. 7 inches
9. 3 inches

Check student answers to verify accurate measurements.

10. How much longer is line 6 than line 7? **7 cm**
11. How much longer is line 8 than line 9? **4 in.**

 73

Name ______________ (2.MD.B.5, 2.MD.B.6)

Relating Addition and Subtraction to Length

A number line is like a ruler. You can use a number line to add or subtract lengths in a word problem.

Mia used 3 inches of blue yarn and 4 inches of red yarn to make a bracelet. How many inches of yarn did Mia use all together?

blue yarn
red yarn
0 1 2 3 4 5 6 7 8 9 10 11 12 13 14 15 16 17 18 19 20

Seven inches of yarn were used altogether.

Jake had 12 yards of rope. He gave his friend 7 yards. How many yards of rope does Jake have now?

Rope Jake had
Rope Jake gave away
0 1 2 3 4 5 6 7 8 9 10 11 12 13 14 15 16 17 18 19 20

Jake has 5 yards of rope left.

Solve the word problems using the number line.

1. Kelly had 17 feet of ribbon. She gave Chris 6 feet. How much ribbon does she have left?
   0 1 2 3 4 5 6 7 8 9 10 11 12 13 14 15 16 17 18 19 20 **11** feet of ribbon
2. Paul had 18 meters of fishing line. Nine meters broke. How many meters are left?
   0 1 2 3 4 5 6 7 8 9 10 11 12 13 14 15 16 17 18 19 20 **9** meters
3. Dante has 8 yards of kite string. He needs 12 more yards. How many yards does he need altogether?
   0 1 2 3 4 5 6 7 8 9 10 11 12 13 14 15 16 17 18 19 20 **20** yards

74

Name ______________ (2.MD.B.5, 2.MD.B.6)

Relating Addition and Subtraction to Length

Draw a number line to solve the problems.

1. Cole had 27 feet of wire. He gave Dante 8 feet. How much wire does Cole have left now?
   0 1 2 3 4 5 6 7 8 9 10 11 12 13 14 15 16 17 18 19 20 21 22 23 24 25 26 27 28 29 30
   19 feet of wire
2. Emily had 41 yards of string to fly a kite. The string broke and Emily had only 20 yards of string left. How many yards of string broke?
   Check students' number lines.
   11 yards
3. Monica has a 14-inch piece of trim to put on a dress. She needs 12 more inches of trim to finish the dress. How many inches does she need altogether?
   26 inches
4. Before Owen sharpened his pencil, it was 16 centimeters long. After he sharpened it, it was 14 centimeters. How many centimeters longer was the pencil before he sharpened it?
   2 centimeters
5. Olivia walked 27 meters on Saturday. Mason walked 15 more meters than Olivia on Sunday. How many meters did Mason walk on Sunday?
   42 meters

 75

Name ______________ (2.MD.B.5, 2.MD.B.6)

Relating Addition and Subtraction to Length

Solve each word problem. Show your work with number lines, number sentences, pictures, or words.

1. Laura had 5 inches of her hair cut. Now, her hair is 14 inches long. How long was Laura's hair before the haircut?
   19 inches
2. Kenneth bought a new 25-foot hose. It is 10 feet longer than his old hose. How many feet long was Kenneth's old hose?
   15 feet
3. Grace needs 42 yards of yarn to make a scarf. She has already used 24 yards. How many more yards does Grace need to finish her scarf?
   18 yards
4. Libby is 12 inches taller than her friend. Her friend is 45 inches tall. How tall is Libby?
   57 inches
5. Jane needs 62 feet of rope to make a swing. She has 42 feet of rope. How much more rope does Jane need to make a swing?
   20 feet
6. Neil saw a 250-foot tall tree in the rain forest. Another tree was 125 feet tall. How many more feet tall is the taller tree?
   125 feet

76

Answer Key

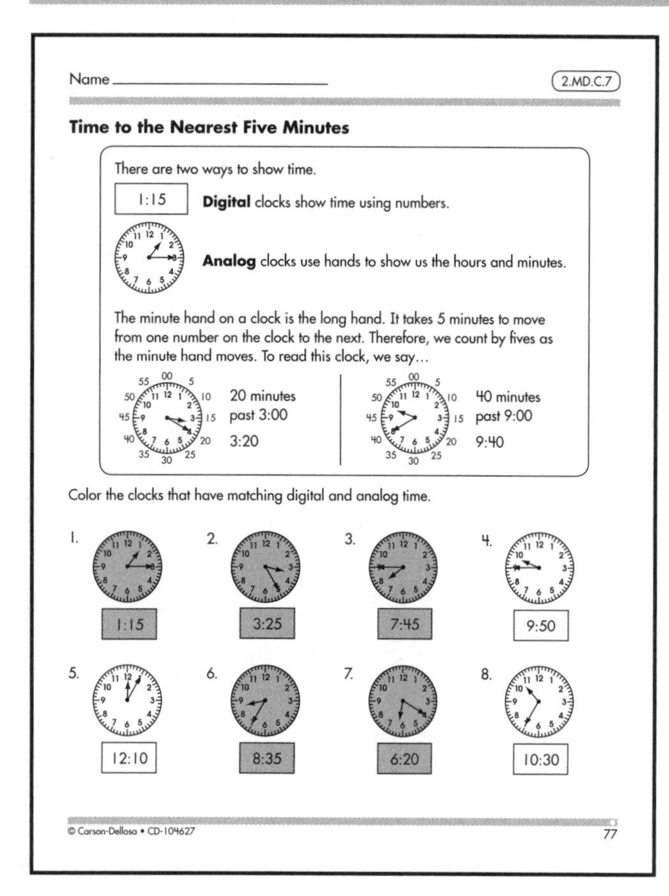

Name ______________________ 2.MD.C.7

Time to the Nearest Five Minutes

There are two ways to show time.

1:15 **Digital** clocks show time using numbers.

Analog clocks use hands to show us the hours and minutes.

The minute hand on a clock is the long hand. It takes 5 minutes to move from one number on the clock to the next. Therefore, we count by fives as the minute hand moves. To read this clock, we say…

20 minutes past 3:00 3:20

40 minutes past 9:00 9:40

Color the clocks that have matching digital and analog time.

1. 1:15 2. 3:25 3. 7:45 4. 9:50

5. 12:10 6. 8:35 7. 6:20 8. 10:30

© Carson-Dellosa • CD-104627 77

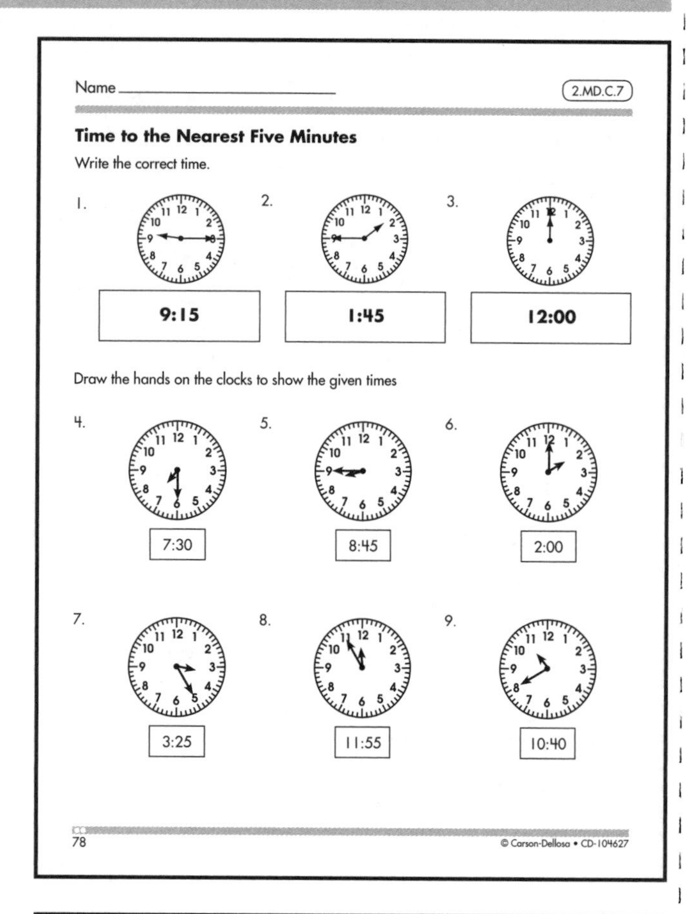

Name ______________________ 2.MD.C.7

Time to the Nearest Five Minutes

Write the correct time.

1. 9:15 2. 1:45 3. 12:00

Draw the hands on the clocks to show the given times

4. 7:30 5. 8:45 6. 2:00

7. 3:25 8. 11:55 9. 10:40

78 © Carson-Dellosa • CD-104627

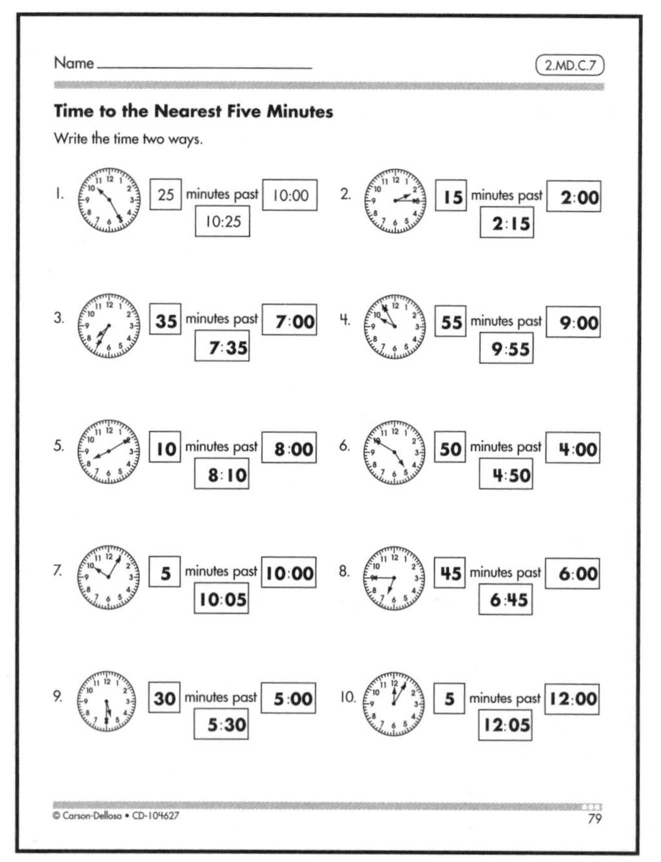

Name ______________________ 2.MD.C.7

Time to the Nearest Five Minutes

Write the time two ways.

1. 25 minutes past 10:00 10:25
2. **15** minutes past **2:00** **2:15**
3. **35** minutes past **7:00** **7:35**
4. **55** minutes past **9:00** **9:55**
5. **10** minutes past **8:00** **8:10**
6. **50** minutes past **4:00** **4:50**
7. **5** minutes past **10:00** **10:05**
8. **45** minutes past **6:00** **6:45**
9. **30** minutes past **5:00** **5:30**
10. **5** minutes past **12:00** **12:05**

© Carson-Dellosa • CD-104627 79

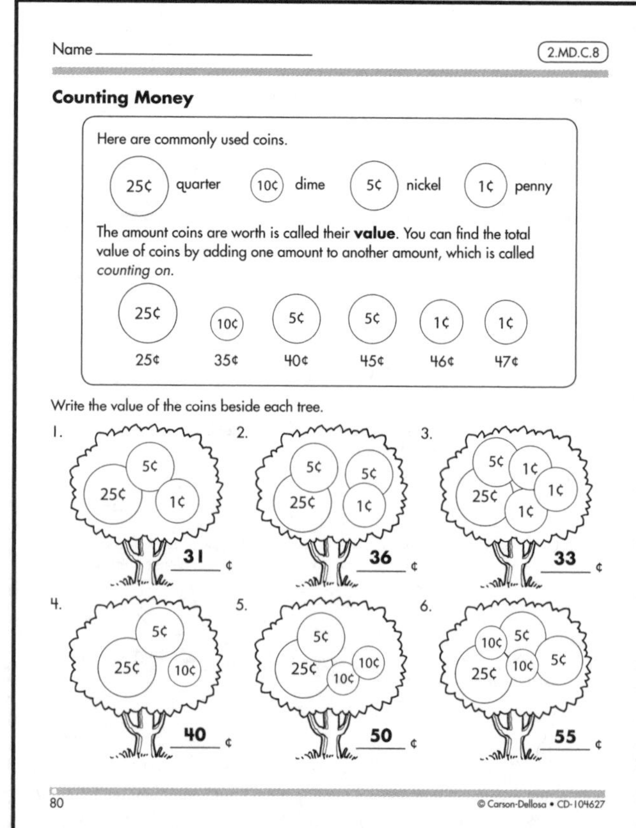

Name ______________________ 2.MD.C.8

Counting Money

Here are commonly used coins.

25¢ quarter 10¢ dime 5¢ nickel 1¢ penny

The amount coins are worth is called their **value**. You can find the total value of coins by adding one amount to another amount, which is called *counting on.*

25¢ 10¢ 5¢ 5¢ 1¢ 1¢

25¢ 35¢ 40¢ 45¢ 46¢ 47¢

Write the value of the coins beside each tree.

1. 25¢ 5¢ 1¢ — **31** ¢
2. 25¢ 5¢ 5¢ 1¢ — **36** ¢
3. 25¢ 5¢ 1¢ 1¢ 1¢ — **33** ¢
4. 25¢ 5¢ 10¢ — **40** ¢
5. 25¢ 5¢ 10¢ 10¢ — **50** ¢
6. 25¢ 10¢ 5¢ 10¢ 5¢ — **55** ¢

80 © Carson-Dellosa • CD-104627

Answer Key

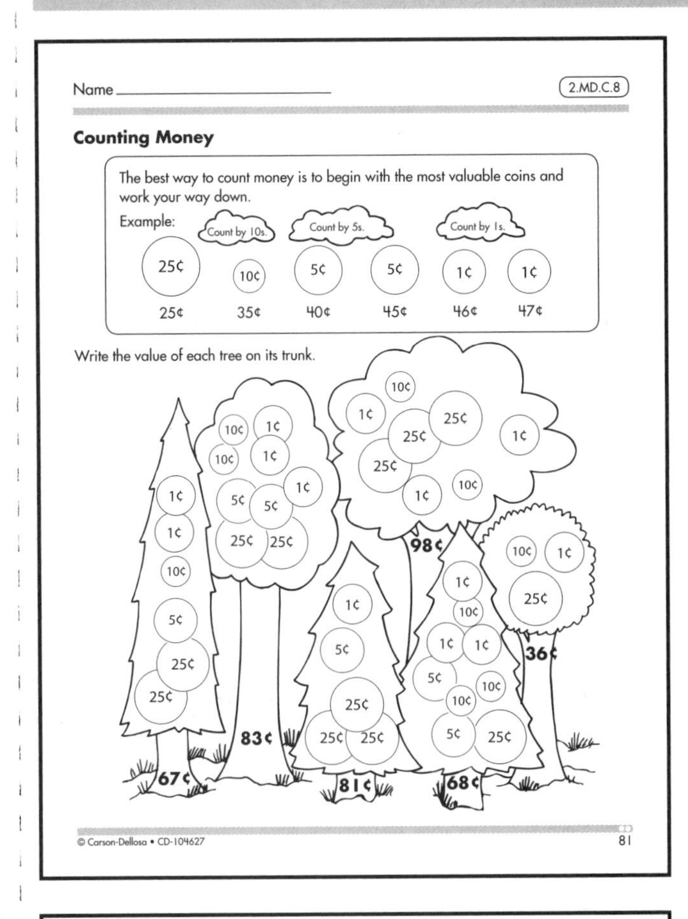

Name ______________________ (2.MD.C.8)

Counting Money

The best way to count money is to begin with the most valuable coins and work your way down.

Example: Count by 10s. Count by 5s. Count by 1s.

25¢ 10¢ 5¢ 5¢ 1¢ 1¢

25¢ 35¢ 40¢ 45¢ 46¢ 47¢

Write the value of each tree on its trunk.

67¢ **83¢** **98¢** **81¢** **68¢** **36¢**

© Carson-Dellosa • CD-104627 81

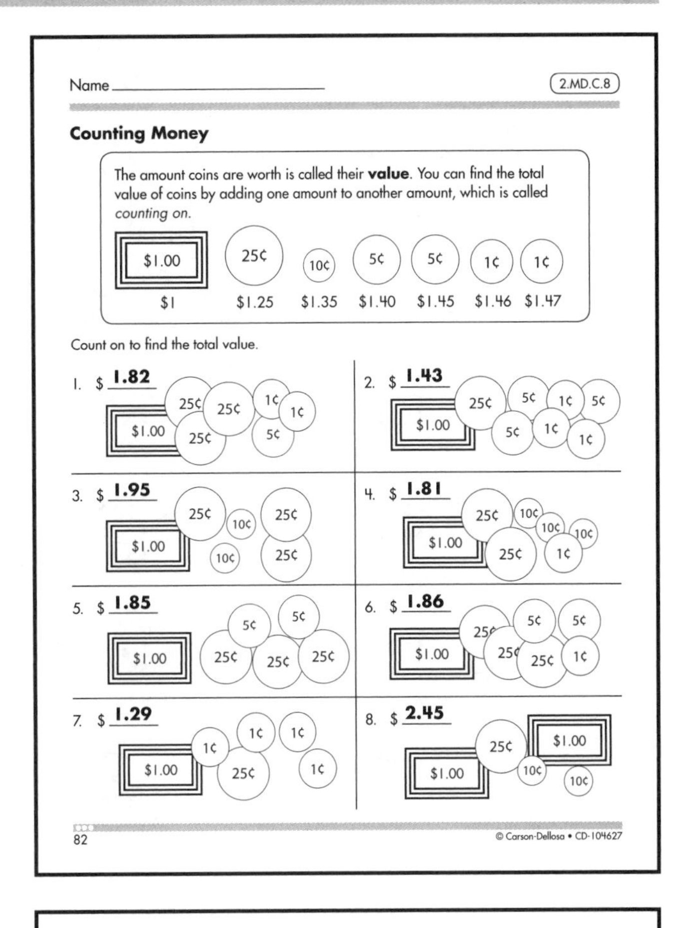

Name ______________________ (2.MD.C.8)

Counting Money

The amount coins are worth is called their **value**. You can find the total value of coins by adding one amount to another amount, which is called *counting on.*

$1.00 25¢ 10¢ 5¢ 5¢ 1¢ 1¢

$1 $1.25 $1.35 $1.40 $1.45 $1.46 $1.47

Count on to find the total value.

1. $ **1.82**
2. $ **1.43**
3. $ **1.95**
4. $ **1.81**
5. $ **1.85**
6. $ **1.86**
7. $ **1.29**
8. $ **2.45**

82 © Carson-Dellosa • CD-104627

Name ______________________ (2.MD.C.8)

Money Problem-Solving

Use the information from the story to solve the math problem. Read each story carefully and decide if you should add or subtract to find the answer.

Add or subtract to find the answers.

1. Nell has 9 pennies. She gives 1 to Judy and 3 to Rick.

   How many pennies does she have left? **5**

   What is their value? **5¢**

2. Hailey has 12 dimes. She gives 5 to Blake and 3 to Jo. Dan gives her 4 more.

   How many dimes does Hailey have now? **8**

   What is their value? **80¢**

3. David has 5 nickels. He finds 6 more.

   How many nickels does David have now? **11**

   What is their value? **55¢**

4. Jessie has 16 pennies. She gives Jeni 9 pennies and then finds 2 more.

   How many pennies does she have in all? **9**

   What is their value? **9¢**

5. Sydney has 8 nickels. Jordi has 5 nickels.

   How many more nickels does Sydney have? **3**

   What is their value? **15¢**

6. Hector has 13 dimes. Meg has 9 dimes.

   How many more dimes does Hector have? **4**

   What is their value? **40¢**

© Carson-Dellosa • CD-104627 83

Name ______________________ (2.MD.C.8)

Money Problem-Solving

Read. Draw the coins. Show more than one way if you can.

Make 40¢.	Make 85¢.	Make 53¢.
1. Taka has 1 quarter. What could the other coins be?	2. Fita has 3 quarters. What could the other coins be? **Answers will vary.**	3. Lara has 6 nickels. What could the other coins be?
4. Kasha has 3 dimes. What could the other coins be?	5. Reba has 4 dimes. What could the other coins be?	6. Rico has 4 dimes. What could the other coins be?
7. Olivia has no dimes or pennies. What could the coins be?	8. Dillon has no nickels or pennies. What could the coins be?	9. Theodore has no dimes. What could the coins be?

84 © Carson-Dellosa • CD-104627

Answer Key

Name ________________ (2.MD.C.8)

Money Problem Solving

Use information from the menu to make a math problem. Solve the problem. Read each story carefully and decide if it makes sense to add or subtract.

Smokey Joe's Barbecue

MAIN DISHES	SIDE DISHES	BEVERAGES
Eye-Watering Ham........$3.50	Flame Fries........$1.10	Cola........$0.75
Burning-Hot Ribs........$3.75	Sizzlin' Salad........$1.05	Lemonade........$0.85
Rockin' Roast Beef........$4.25	Tasty Tater Tots........$0.95	Milk........$0.95

1. Ariel ordered ribs and lemonade. How much will her lunch cost? **3.75 + .85 = $4.60**
2. Michael ordered roast beef. He paid with a five-dollar bill. How much change will he get? **5.00 − 4.25 = $0.75**
3. Jonah has $4.08. He buys ham as a main dish. How much money does Jonah have left? **4.08 − 3.50 = $0.58**
4. Terone wonders, "How much does an order of ribs, fries, and a cola cost?" **3.75 + 1.10 + .75 = $5.60**
5. How much more does roast beef cost than milk? **4.25 − .95 = $3.30**
6. Kelsey orders the least expensive item from each section of the menu. How much does she spend? **3.50 + .95 + .75 = $5.20**
7. Ryan spent $5.55 for lunch. He got $0.45 back as change. How much money did Ryan start out with? **5.55 + .45 = $6.00**
8. Tracy buys lemonade for herself and three friends. How much does she spend? **.85 + .85 + .85 + .85 = $3.40**

Name ________________ (2.MD.D.9)

Line Plots

A line plot uses a number line and Xs to show data that has been collected.

The line plot shows the height of the sunflowers in Ms. Park's garden. Read the graph and answer the questions.

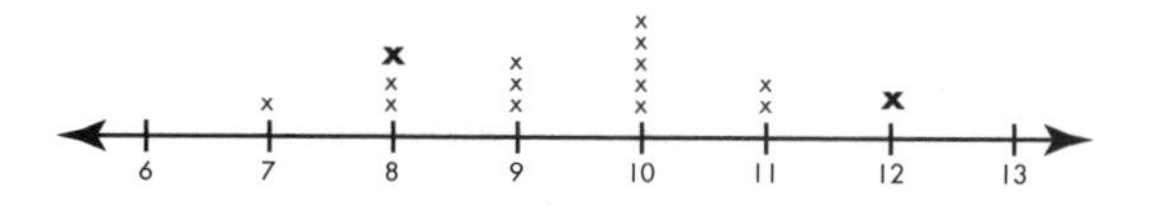

1. What is the most common height of the sunflowers?
   10 ft.
2. How many sunflowers are 10 feet tall?
   5 sunflowers
3. How many sunflowers are 8 feet tall?
   2 sunflowers
4. Which height shows three sunflowers?
   9 ft.
5. Ms. Park measured two more sunflowers. The first one was 8 feet tall and the second one was 12 feet tall. Mark **X**s on the number line to plot the sunflowers on the graph.

Name ________________ (2.MD.D.9)

Line Plots

Brooke made necklaces of different lengths to sell at the school carnival.

Lengths of Necklaces

16 inches	18 inches
20 inches	17 inches
16 inches	16 inches
17 inches	19 inches
18 inches	16 inches

Use the data to complete the line plot. Answer the questions.

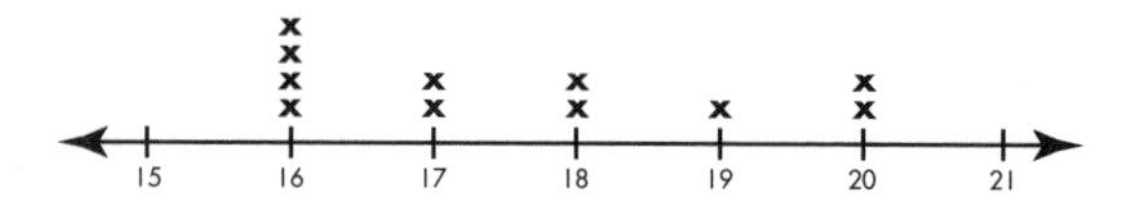

1. What was the total number of 16-inch necklaces?
   4 necklaces
2. What was the total number of 18-inch necklaces?
   2 necklaces
3. Brooke made one more necklace that was 20 inches long. Graph that necklace on the line plot.
4. What was the total number of necklaces that Brooke made?
   11 necklaces

Name ________________ (2.MD.D.9)

Line Plots

Mrs. Rivera's class planted beans. After a week, the class recorded the height of the sprouts. Use the following data to make a line plot in the space provided. Answer the questions.

Bean Sprouts

Height in centimeters	Number they counted
3 cm	4
4 cm	7
5 cm	6
6 cm	7
7 cm	3

Check students' work.

1. Which two heights had the same number of bean sprouts?
   4 cm, 6 cm
2. How many bean sprouts were 7 centimeters in height?
   3 bean sprouts
3. What was the total number of bean sprouts?
   27 bean sprouts
4. How many more bean sprouts were 5 centimeters than 3 centimeters in height?
   2 bean sprouts

Answer Key

Name ______________________ (2.MD.D.10)

Interpreting Graphs

Graphs can be used to observe and compare information. **Picture graphs**, or **pictographs**, often have a key that will tell what each picture means. In this graph, one circle equals two students.

Margie's class made a frequency table to show what the second graders did during choice time. Use the table to make a pictograph below.

Activity	Number of Students
read	18
finish work	6
dice math	24
color	12
clean desk	6
science project	18

Draw 1 ○ for every 2 students.

Activity	Number of Students
read	○○○○○○○○○
finish work	○○○
dice math	○○○○○○○○○○○○
color	○○○○○○
clean desk	○○○
science project	○○○○○○○○○

○ = 2 students

1. If 10 students made an art project, how many circles would you draw? **5 circles**

2. If each ○ = 2 students, how would you show 1 student? ◖ 9 students? ○○○○◖

3. Circle **T** for true or **F** for false.

   (T) F More students chose dice math than coloring.
   (T) F An equal number of students read as did the science project.
   T (F) Fewer students colored than cleaned desks.
   T (F) Ten more students finished work than read.

© Carson-Dellosa • CD-104627 89

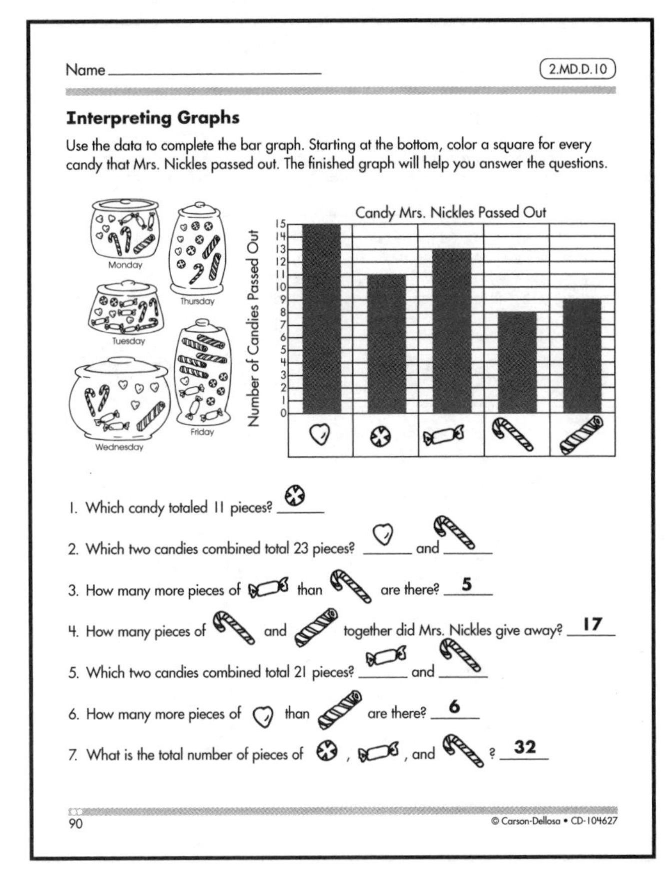

Name ______________________ (2.MD.D.10)

Interpreting Graphs

Use the data to complete the bar graph. Starting at the bottom, color a square for every candy that Mrs. Nickles passed out. The finished graph will help you answer the questions.

1. Which candy totaled 11 pieces?
2. Which two candies combined total 23 pieces? ____ and ____
3. How many more pieces of ____ than ____ are there? **5**
4. How many pieces of ____ and ____ together did Mrs. Nickles give away? **17**
5. Which two candies combined total 21 pieces? ____ and ____
6. How many more pieces of ____ than ____ are there? **6**
7. What is the total number of pieces of ____, ____, and ____? **32**

90 © Carson-Dellosa • CD-104627

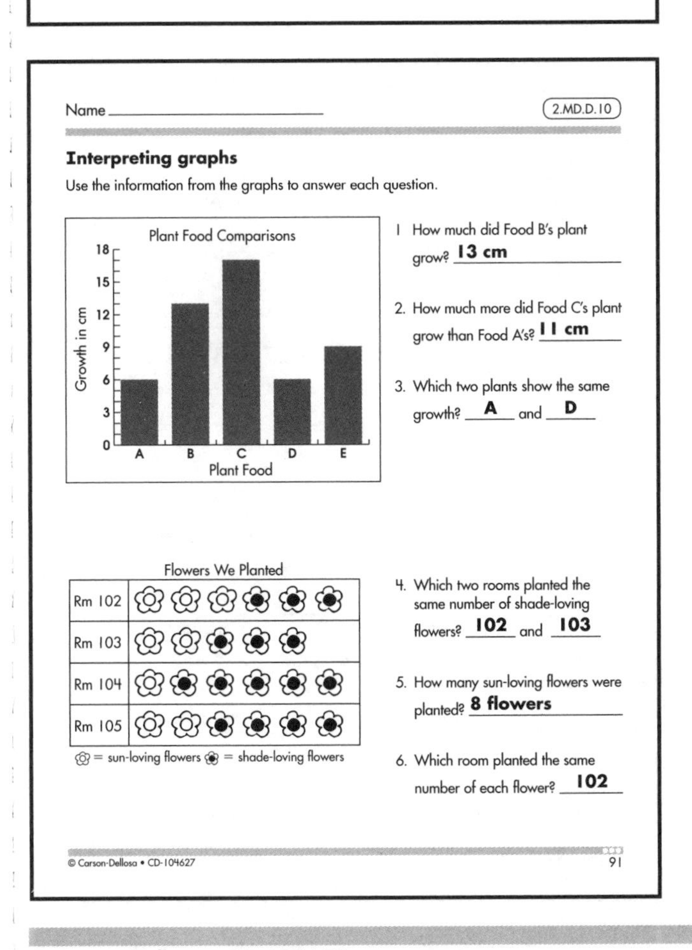

Name ______________________ (2.MD.D.10)

Interpreting graphs

Use the information from the graphs to answer each question.

1. How much did Food B's plant grow? **13 cm**
2. How much more did Food C's plant grow than Food A's? **11 cm**
3. Which two plants show the same growth? **A** and **D**

4. Which two rooms planted the same number of shade-loving flowers? **102** and **103**
5. How many sun-loving flowers were planted? **8 flowers**
6. Which room planted the same number of each flower? **102**

© Carson-Dellosa • CD-104627 91

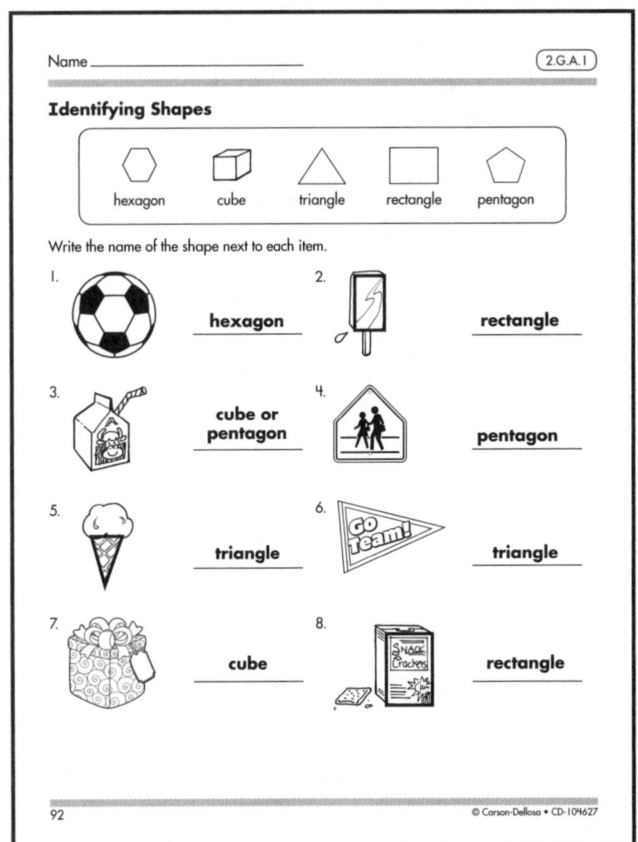

Name ______________________ (2.G.A.1)

Identifying Shapes

hexagon cube triangle rectangle pentagon

Write the name of the shape next to each item.

1. **hexagon**
2. **rectangle**
3. **cube or pentagon**
4. **pentagon**
5. **triangle**
6. **triangle**
7. **cube**
8. **rectangle**

92 © Carson-Dellosa • CD-104627

Answer Key

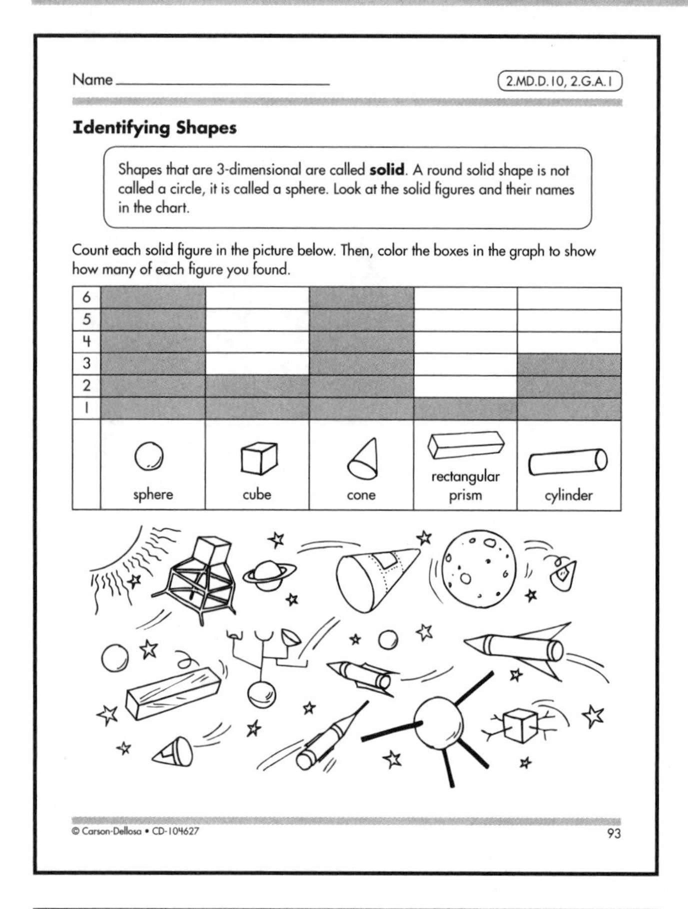

Name ____________________ (2.MD.D.10, 2.G.A.1)

Identifying Shapes

Shapes that are 3-dimensional are called **solid**. A round solid shape is not called a circle, it is called a sphere. Look at the solid figures and their names in the chart.

Count each solid figure in the picture below. Then, color the boxes in the graph to show how many of each figure you found.

6
5
4
3
2
1

sphere | cube | cone | rectangular prism | cylinder

© Carson-Dellosa • CD-104627 93

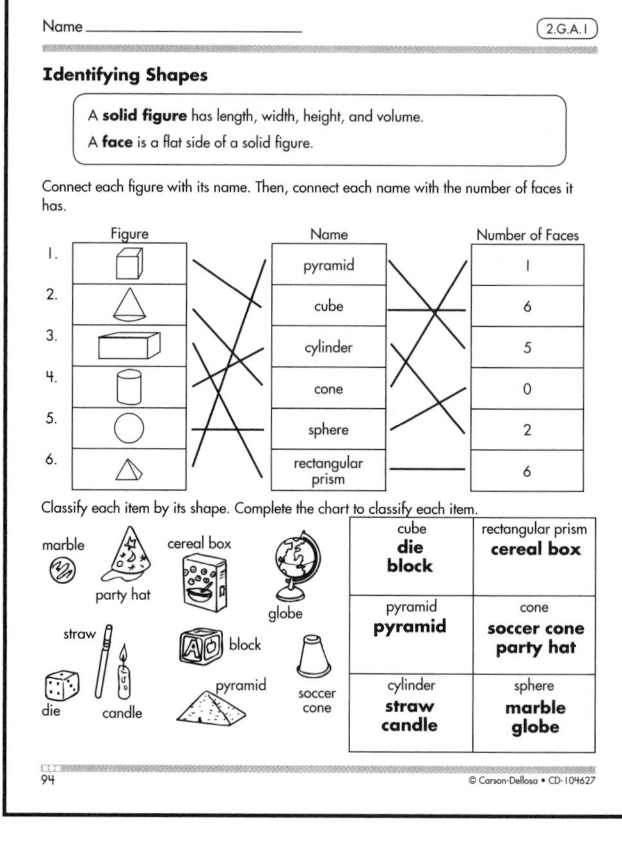

Name ____________________ (2.G.A.1)

Identifying Shapes

A **solid figure** has length, width, height, and volume.
A **face** is a flat side of a solid figure.

Connect each figure with its name. Then, connect each name with the number of faces it has.

	Figure	Name	Number of Faces
1.		pyramid	1
2.		cube	6
3.		cylinder	5
4.		cone	0
5.		sphere	2
6.		rectangular prism	6

Classify each item by its shape. Complete the chart to classify each item.

marble, party hat, cereal box, globe, straw, block, die, candle, pyramid, soccer cone

cube **die** **block**	rectangular prism **cereal box**
pyramid **pyramid**	cone **soccer cone** **party hat**
cylinder **straw** **candle**	sphere **marble** **globe**

94 © Carson-Dellosa • CD-104627

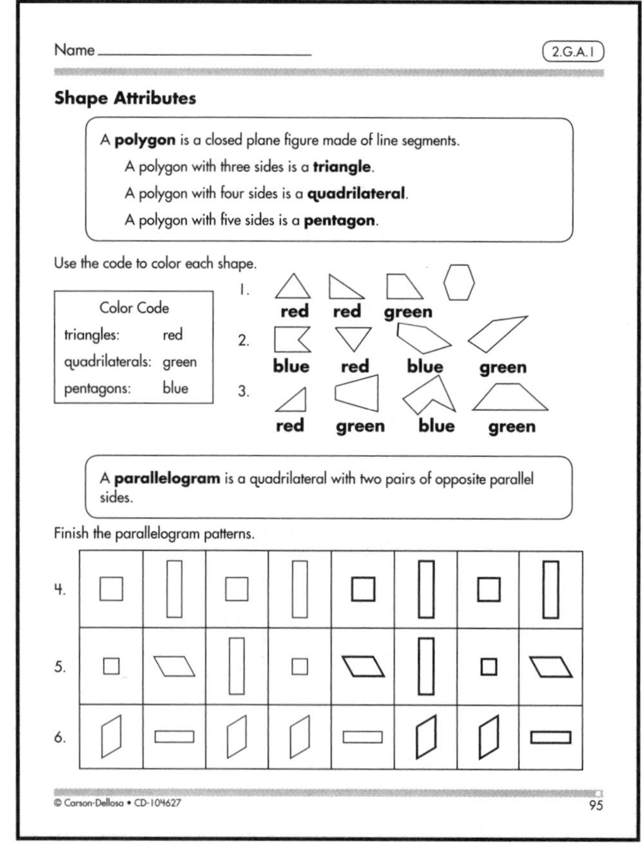

Name ____________________ (2.G.A.1)

Shape Attributes

A **polygon** is a closed plane figure made of line segments.
A polygon with three sides is a **triangle**.
A polygon with four sides is a **quadrilateral**.
A polygon with five sides is a **pentagon**.

Use the code to color each shape.

Color Code	
triangles:	red
quadrilaterals:	green
pentagons:	blue

1. **red** **red** **green**
2. **blue** **red** **blue** **green**
3. **red** **green** **blue** **green**

A **parallelogram** is a quadrilateral with two pairs of opposite parallel sides.

Finish the parallelogram patterns.

4.
5.
6.

© Carson-Dellosa • CD-104627 95

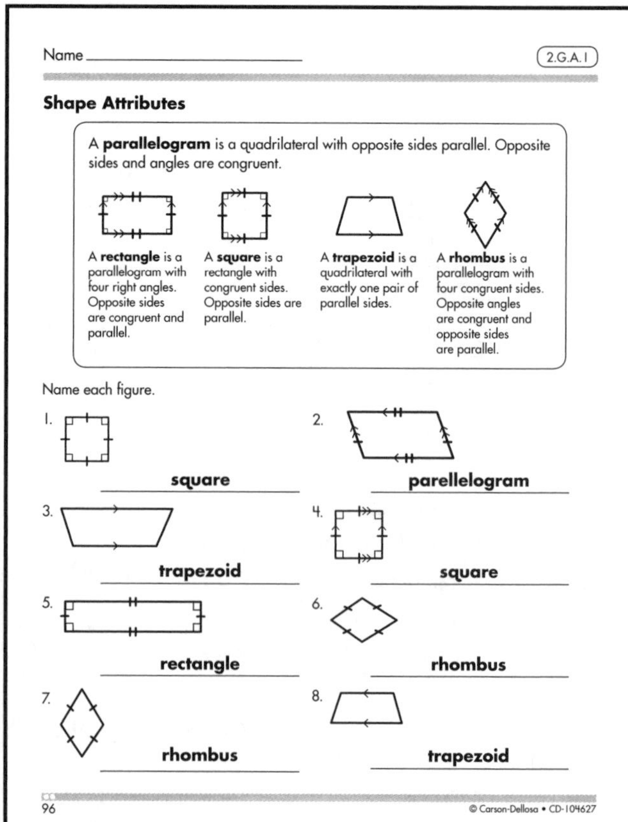

Name ____________________ (2.G.A.1)

Shape Attributes

A **parallelogram** is a quadrilateral with opposite sides parallel. Opposite sides and angles are congruent.

A **rectangle** is a parallelogram with four right angles. Opposite sides are congruent and parallel.

A **square** is a rectangle with congruent sides. Opposite sides are parallel.

A **trapezoid** is a quadrilateral with exactly one pair of parallel sides.

A **rhombus** is a parallelogram with four congruent sides. Opposite angles are congruent and opposite sides are parallel.

Name each figure.

1. **square**
2. **parellelogram**
3. **trapezoid**
4. **square**
5. **rectangle**
6. **rhombus**
7. **rhombus**
8. **trapezoid**

96 © Carson-Dellosa • CD-104627

Answer Key

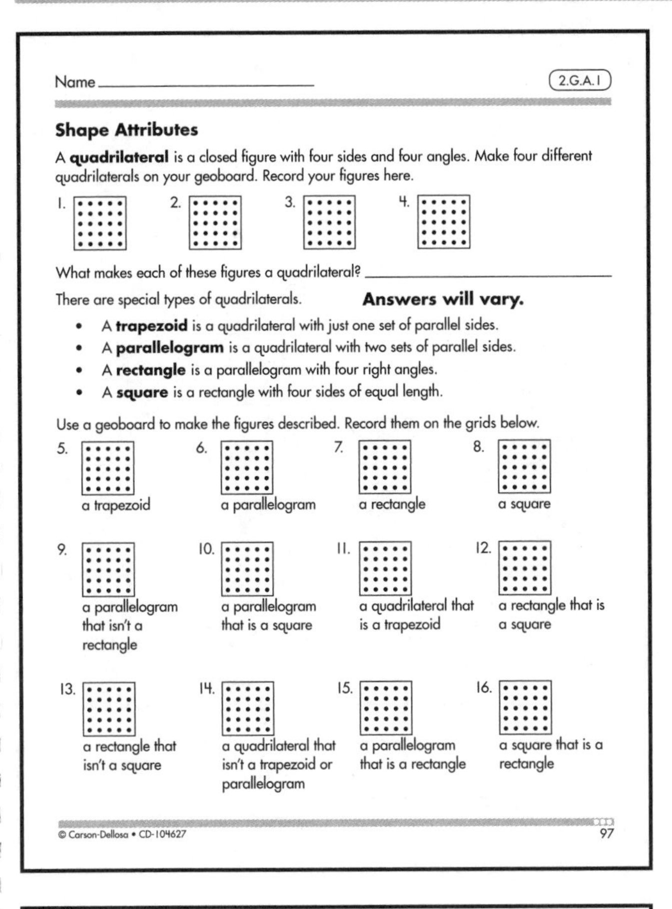

Name ______________________ (2.G.A.1)

Shape Attributes

A **quadrilateral** is a closed figure with four sides and four angles. Make four different quadrilaterals on your geoboard. Record your figures here.

1. 2. 3. 4.

What makes each of these figures a quadrilateral? ______________________

There are special types of quadrilaterals. **Answers will vary.**

- A **trapezoid** is a quadrilateral with just one set of parallel sides.
- A **parallelogram** is a quadrilateral with two sets of parallel sides.
- A **rectangle** is a parallelogram with four right angles.
- A **square** is a rectangle with four sides of equal length.

Use a geoboard to make the figures described. Record them on the grids below.

5. a trapezoid
6. a parallelogram
7. a rectangle
8. a square
9. a parallelogram that isn't a rectangle
10. a parallelogram that is a square
11. a quadrilateral that is a trapezoid
12. a rectangle that is a square
13. a rectangle that isn't a square
14. a quadrilateral that isn't a trapezoid or parallelogram
15. a parallelogram that is a rectangle
16. a square that is a rectangle

© Carson-Dellosa • CD-104627 97

Name ______________________ (2.G.A.2)

Partitioning Rectangles

You can **partition** rectangles into equal parts using rows and columns. This rectangle has 9 equal parts with 3 rows and 3 columns.

1. Number of rows? **5**
   Number of columns? **3**
2. Number of rows? **2**
   Number of columns? **8**
3. Number of rows? **3**
   Number of columns? **4**
4. Partition this rectangle equally into 6 columns and 2 rows.
5. Partition this rectangle equally into 4 rows and 5 columns.
6. Partition this rectangle equally into 5 columns and 5 rows.

Check students' answers to verify that each rectangle is partitioned correctly.

98 © Carson-Dellosa • CD-104627

Name ______________________ (2.G.A.2)

Partitioning Rectangles

Check students' answers to verify that each rectangle is partitioned correctly.

Use the grid to make a rectangle with:

1. 6 rows and 3 columns
2. 1 row and 6 columns
3. 4 rows and 4 columns
4. 5 rows and 3 columns
5. Divide this rectangle into 7 columns and 2 rows.

   How many parts did you draw? **14**
6. Divide this rectangle into 4 rows and 6 columns.

   How many parts did you draw? **24**

© Carson-Dellosa • CD-104627 99

Name ______________________ (2.G.A.2)

Partitioning Rectangles

Using color tiles or square pattern blocks, draw and record as many different rectangles as you can with the given number.

1. 8 squares **Answers will vary.**
2. 10 squares
3. 12 squares

100 © Carson-Dellosa • CD-104627

Answer Key

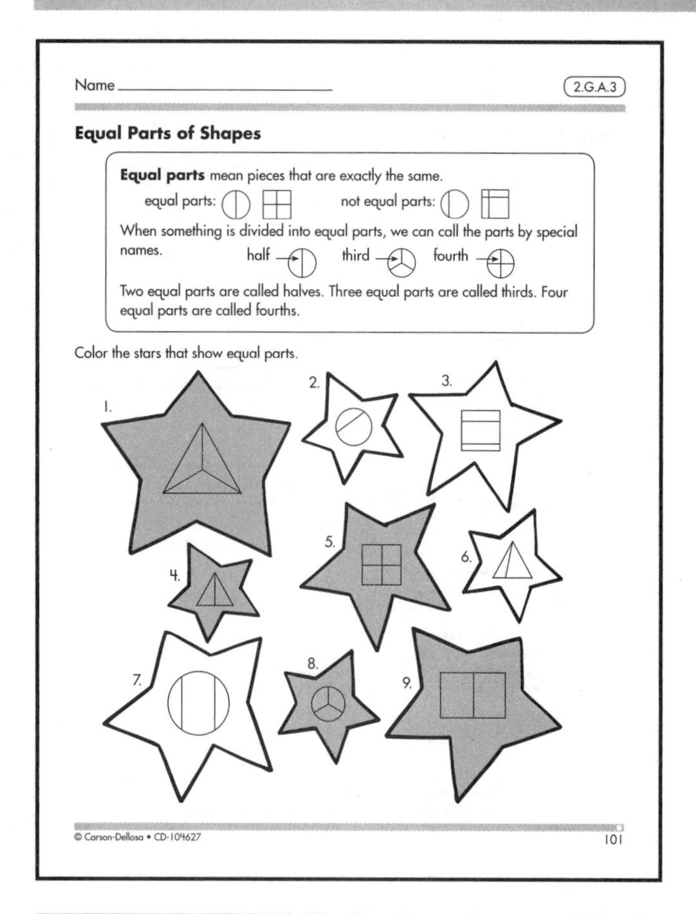
Name ____________________ 2.G.A.3

Equal Parts of Shapes

Equal parts mean pieces that are exactly the same.

equal parts: not equal parts:

When something is divided into equal parts, we can call the parts by special names. half third fourth

Two equal parts are called halves. Three equal parts are called thirds. Four equal parts are called fourths.

Color the stars that show equal parts.

1. 2. 3. 4. 5. 6. 7. 8. 9.

© Carson-Dellosa • CD-104627 101

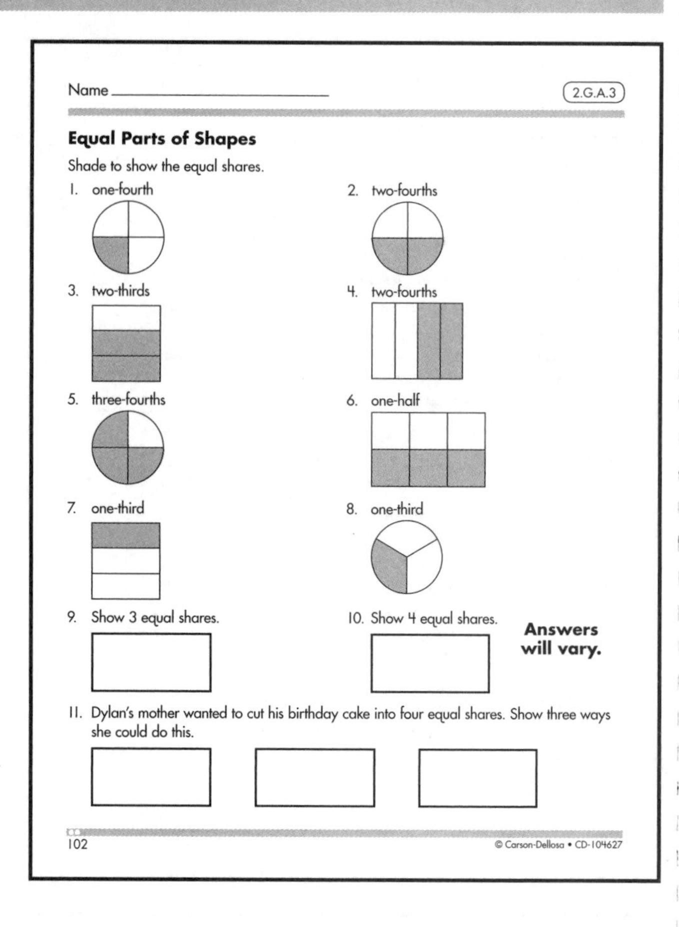
Name ____________________ 2.G.A.3

Equal Parts of Shapes

Shade to show the equal shares.

1. one-fourth
2. two-fourths
3. two-thirds
4. two-fourths
5. three-fourths
6. one-half
7. one-third
8. one-third
9. Show 3 equal shares.
10. Show 4 equal shares.

Answers will vary.

11. Dylan's mother wanted to cut his birthday cake into four equal shares. Show three ways she could do this.

102 © Carson-Dellosa • CD-104627

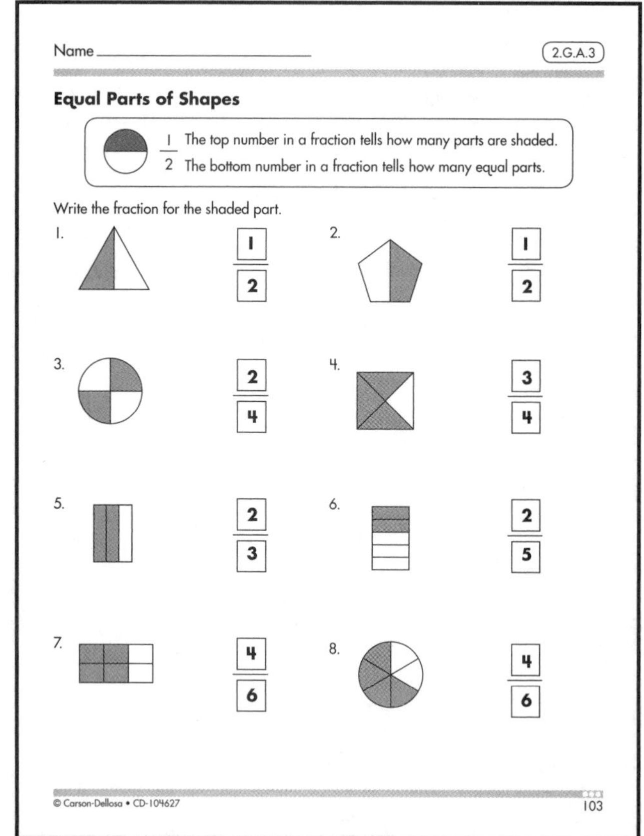
Name ____________________ 2.G.A.3

Equal Parts of Shapes

$\frac{1}{2}$ The top number in a fraction tells how many parts are shaded. The bottom number in a fraction tells how many equal parts.

Write the fraction for the shaded part.

1. $\frac{1}{2}$
2. $\frac{1}{2}$
3. $\frac{2}{4}$
4. $\frac{3}{4}$
5. $\frac{2}{3}$
6. $\frac{2}{5}$
7. $\frac{4}{6}$
8. $\frac{4}{6}$

© Carson-Dellosa • CD-104627 103

Congratulations!

____

receives this award for

____

Signed ________________________________ Date ____________________

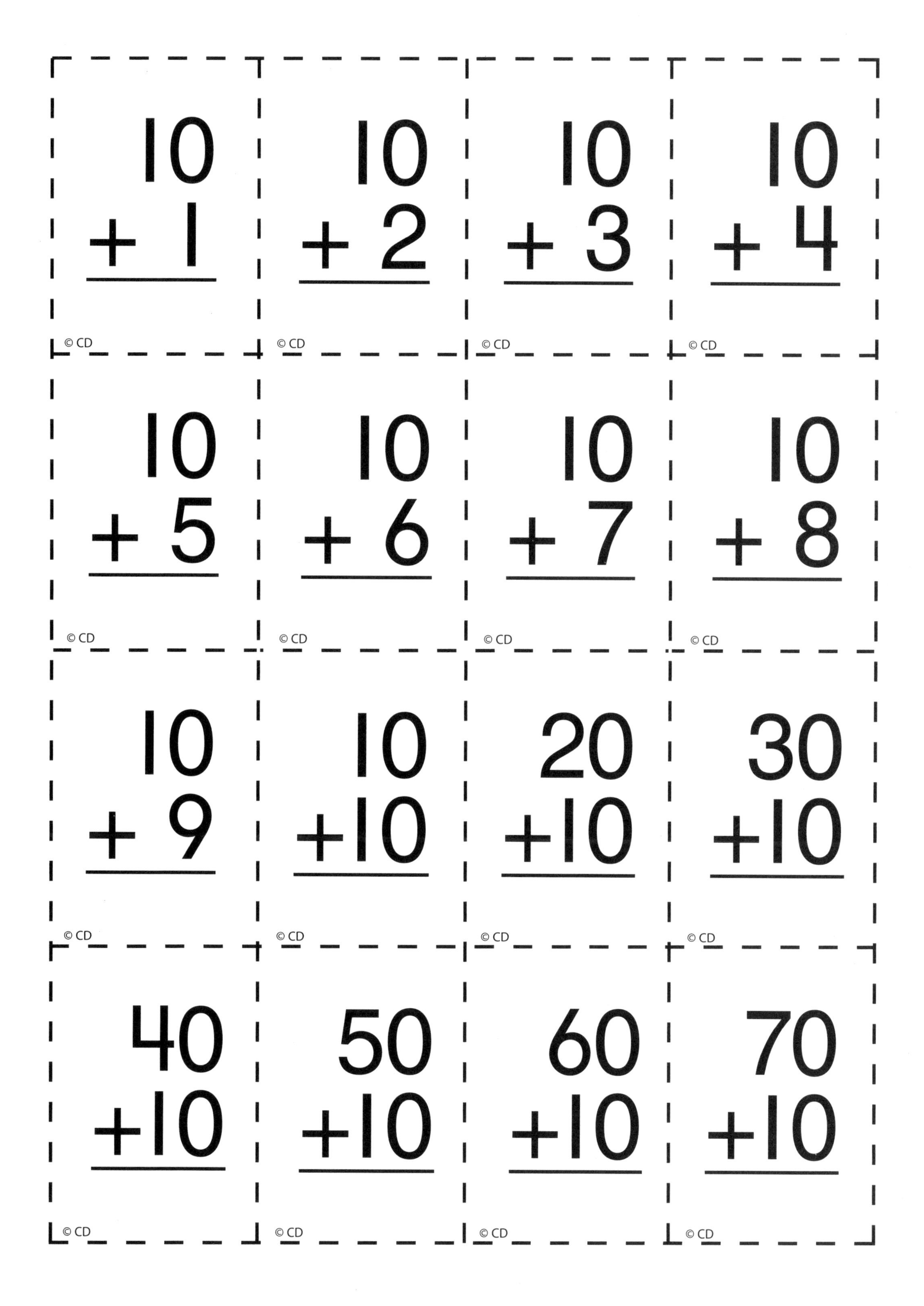
10 + 1
© CD
10 + 2
© CD
10 + 3
© CD
10 + 4
© CD
10 + 5
© CD
10 + 6
© CD
10 + 7
© CD
10 + 8
© CD
10 + 9
© CD
10 + 10
© CD
20 + 10
© CD
30 + 10
© CD
40 + 10
© CD
50 + 10
© CD
60 + 10
© CD
70 + 10
© CD

14	13	12	11
18	17	16	15
40	30	20	19
80	70	60	50

$$\begin{array}{r} 80 \\ +10 \\ \hline \end{array}$$

$$\begin{array}{r} 90 \\ +10 \\ \hline \end{array}$$

$$\begin{array}{r} 100 \\ +10 \\ \hline \end{array}$$

$$\begin{array}{r} 100 \\ +20 \\ \hline \end{array}$$

$$\begin{array}{r} 100 \\ +30 \\ \hline \end{array}$$

$$\begin{array}{r} 100 \\ +40 \\ \hline \end{array}$$

$$\begin{array}{r} 100 \\ +50 \\ \hline \end{array}$$

$$\begin{array}{r} 100 \\ +60 \\ \hline \end{array}$$

$$\begin{array}{r} 100 \\ +70 \\ \hline \end{array}$$

$$\begin{array}{r} 100 \\ +80 \\ \hline \end{array}$$

$$\begin{array}{r} 100 \\ +90 \\ \hline \end{array}$$

$$\begin{array}{r} 100 \\ -\ 10 \\ \hline \end{array}$$

$$\begin{array}{r} 100 \\ -\ 20 \\ \hline \end{array}$$

$$\begin{array}{r} 100 \\ -\ 30 \\ \hline \end{array}$$

$$\begin{array}{r} 100 \\ -\ 40 \\ \hline \end{array}$$

$$\begin{array}{r} 100 \\ -\ 50 \\ \hline \end{array}$$

120	110	100	90
160	150	140	130
90	190	180	170
50	60	70	80

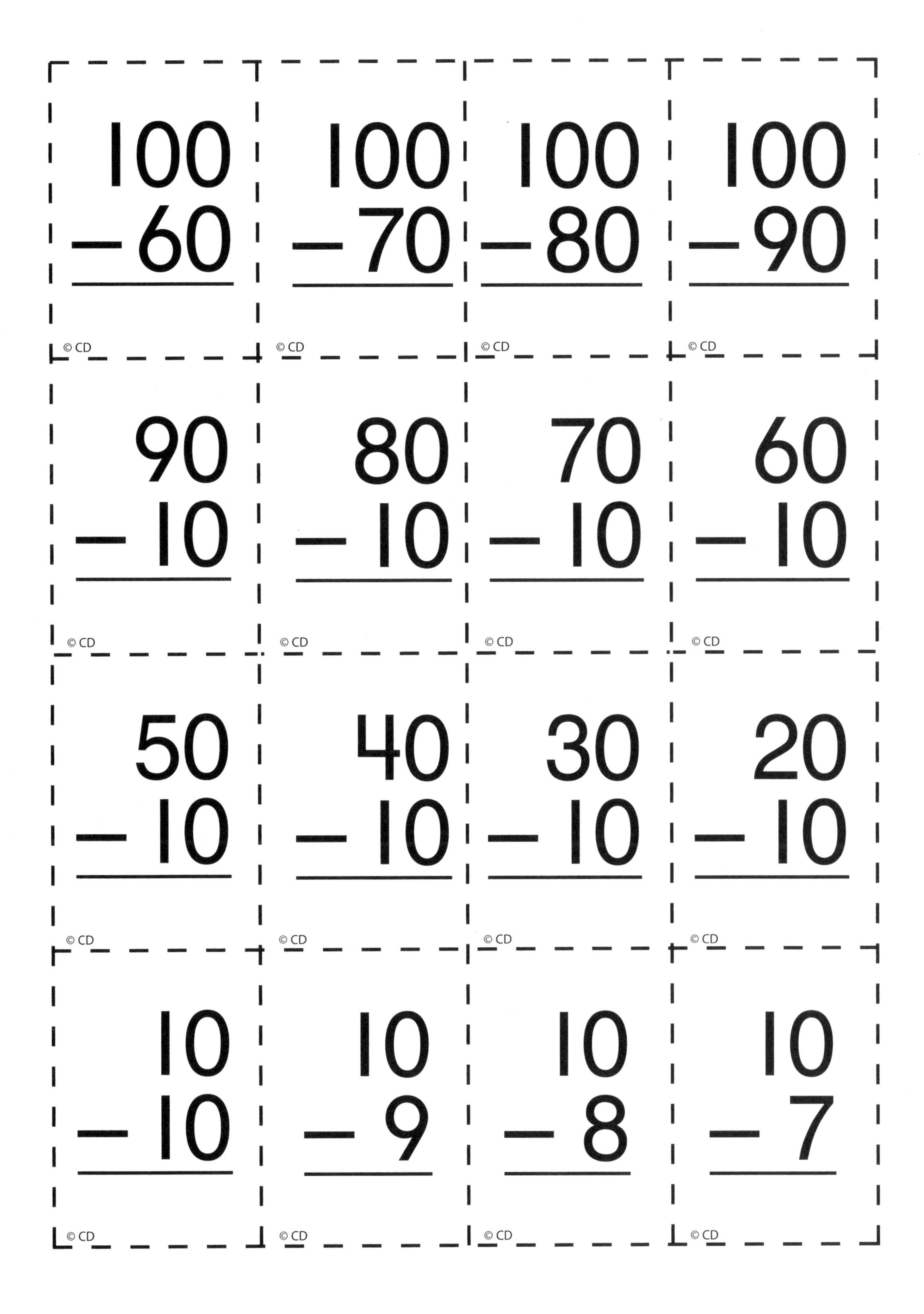
100 − 60
© CD
100 − 70
© CD
100 − 80
© CD
100 − 90
© CD
90 − 10
© CD
80 − 10
© CD
70 − 10
© CD
60 − 10
© CD
50 − 10
© CD
40 − 10
© CD
30 − 10
© CD
20 − 10
© CD
10 − 10
© CD
10 − 9
© CD
10 − 8
© CD
10 − 7
© CD

10	20	30	40
50	60	70	80
10	20	30	40
3	2	1	0

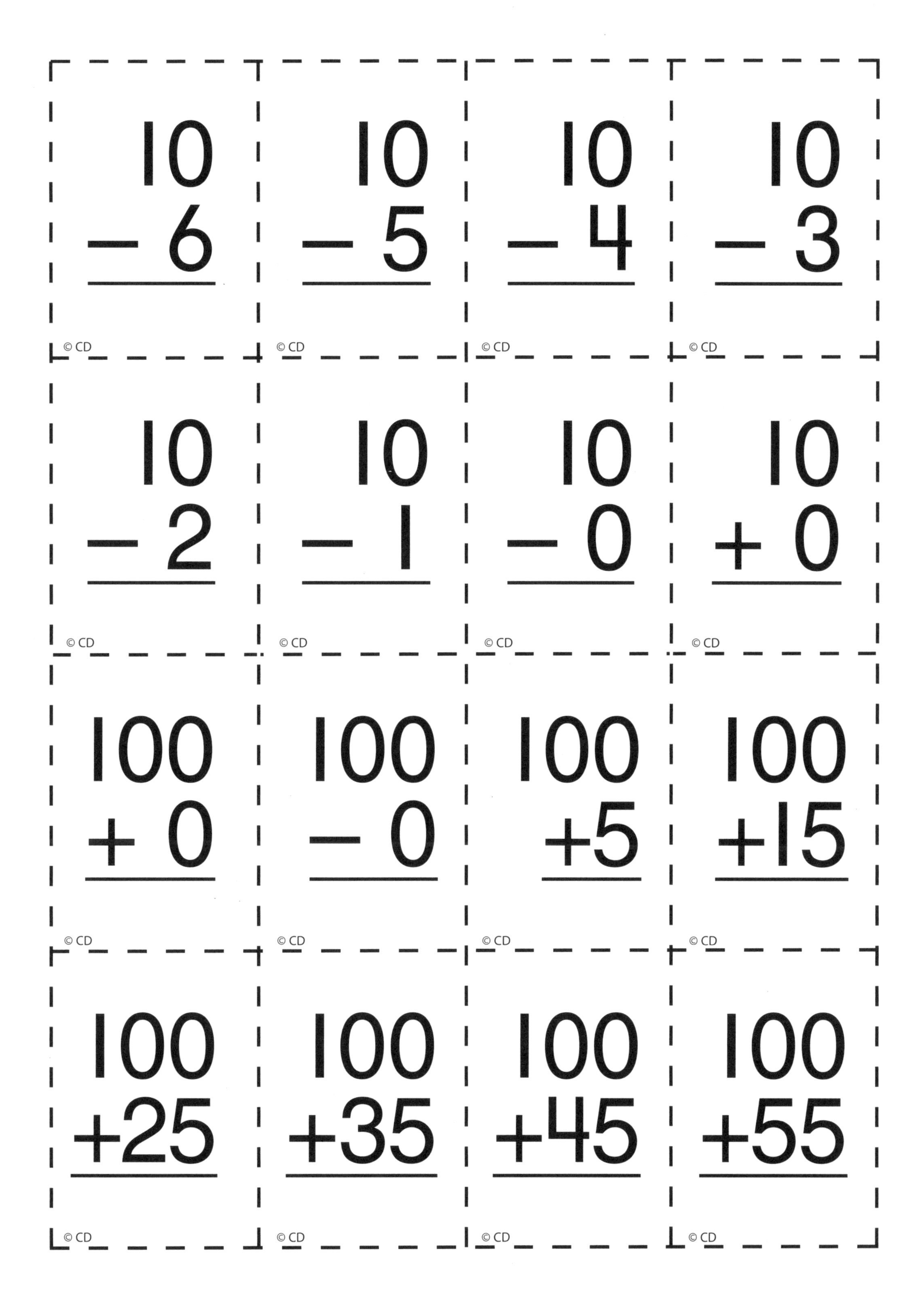
10 − 6
© CD
10 − 5
© CD
10 − 4
© CD
10 − 3
© CD
10 − 2
© CD
10 − 1
© CD
10 − 0
© CD
10 + 0
© CD
100 + 0
© CD
100 − 0
© CD
100 +5
© CD
100 +15
© CD
100 +25
© CD
100 +35
© CD
100 +45
© CD
100 +55
© CD

7 6 5 4

10 10 9 8

115 105 100 100

155 145 135 125

100 +65	100 +75	100 +85	100 +95
95 +10	85 +10	75 +10	65 +10
55 +10	45 +10	35 +10	25 +10
15 +10	200 +25	300 +50	400 +75

195	185	175	165
75	85	95	105
35	45	55	65
475	350	225	25

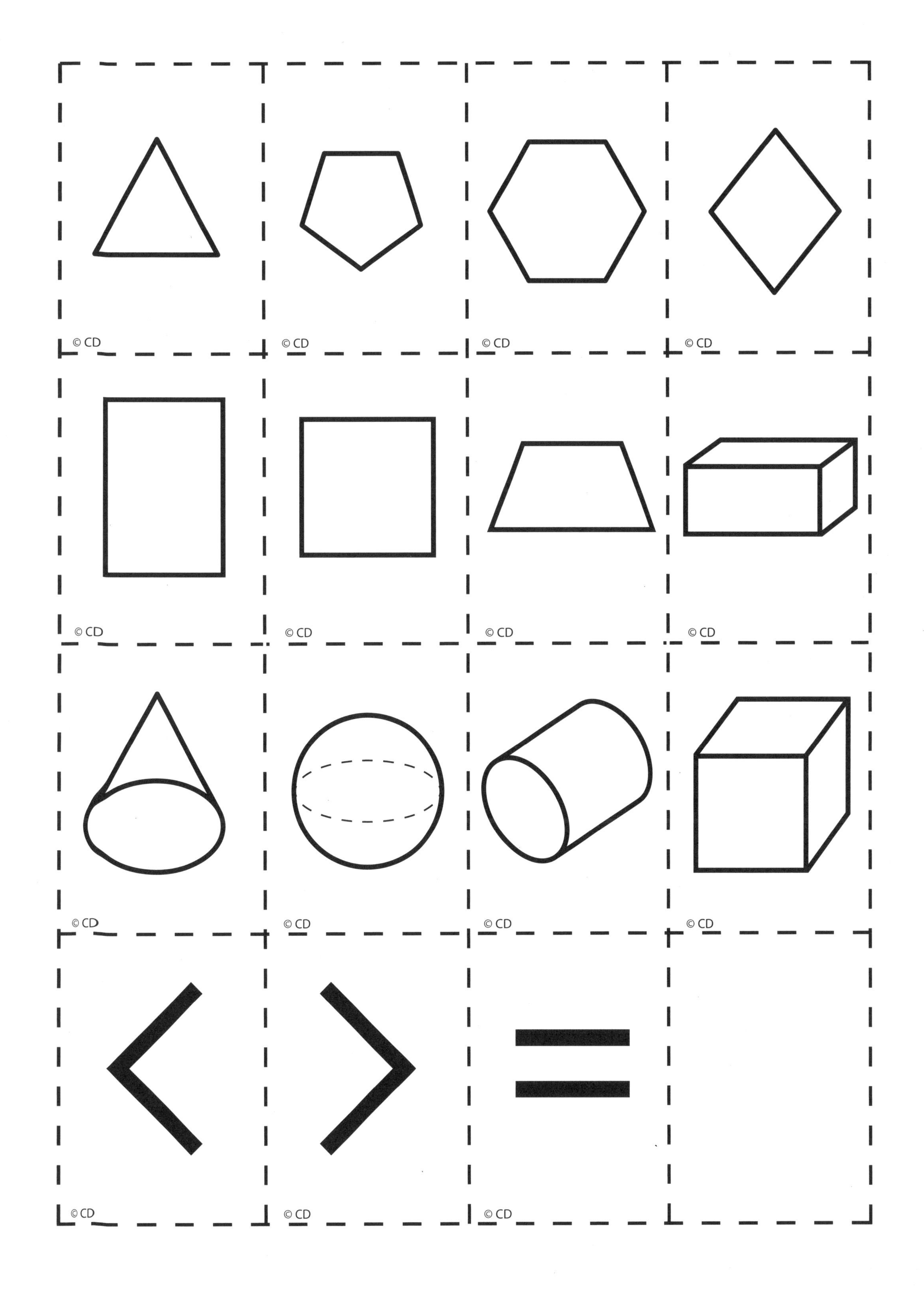
© CD
© CD
© CD
© CD
© CD
© CD
© CD
© CD
© CD
© CD
© CD
© CD
© CD
© CD
© CD

rhombus

hexagon

pentagon

triangle

rectangular prism

trapezoid

square

rectangle

cube

cylinder

sphere

cone

equal to

greater than

less than